The Marauder

&

A Memoir from the 1944 Diary of
MERRILL'S MARAUDER
Larry W. Stephenson

LINDA S. CUNNINGHAM

The Marauder and His Daughter

A Memoir from the 1944 Diary of Merrill's Marauder Larry W. Stephenson

© 2023, Linda S. Cunningham.

All rights reserved. This book or any portion thereof may not be reproduced or used in any manner whatsoever without the express written permission of the publisher except for the use of brief quotations in a book review.

All rights reserved.

First published and printed in the United States of America. No part of this book may be used or reproduced in any manner whatsoever without written permission except in the case of brief quotations embodied in critical articles and reviews.

For information, address **Book Writing Cube**

Head Office: 4145 S.W. Watson, Suite 350, Beaverton, Oregon 97005, United States

Operations: 8 The Green STE 300, Dover DE 19901

https://www.bookwritingcube.com/

Published by **Book Writing Cube**

Printed in the United States

Copyright (c) 2023-24 by Linda S. Cunningham

ISBN Paperback: 978-1-961563-20-9

ISBN Hardcover: 978-1-961563-21-6

Library of Congress Control Number: 2023921690

*For the great men who were the legendary
Merrill's Marauders and especially
my father Marauder,
Larry W. Stephenson
and his family*

"Your subconscious and your body are more in tune with the divine than your mind will ever understand."

Acknowledgments

I am grateful for the input and encouragement from my husband, Patrick Cunningham, and our daughters, Larissa Fahrenholz and Candice Burrows. Their editing and Candice's contribution of memorabilia from the *Lurline* were of tremendous value.

Thanks to the Stephenson family, especially the proofreading and reminiscing with my sister Juanell Stephenson, whose many contributions were immeasurable. Thanks also for the specific memories of my brother Larry W. Stephenson, Jr.

I wish to mention my siblings, Lorise Grimball and Grant Stephenson, who may have been too young to remember specific memories of Dad's PTSD, but they loved their father as well.

Many thanks to my Aunt Miriam Corbello for her memories of my mother during her wartime separation from my father. Thanks to my grandson Douglas Fahrenholz for his knowledge of the Chinese and Japanese languages, histories, and cultures.

I am grateful for Dan Leone's WWII aircraft expertise. Thanks to Colonel Logan Weston's daughter Sherry Cowling with her insight and knowledge of her father's experience. Thanks to Gil and Maxine Darcey, as well as John and Michelle Sorrentino, for their support and input to make the book better.

Thanks also for the input from Jonnie Clasen with the Merrill's Marauder's Proud Descendants Organization.

I am also grateful to Sergeant Warren Porter, Army Ranger historian, for his help in interpreting the diary as well as all things military.

Many thanks for the exceptional work of marketing, editing and layout teams by Book Writing Cube.

Finally, I am grateful to Eben Thomas, whose outstanding editing made this book a reality.

Preface

History is important because it becomes the measuring tool of how much humans have grown or regressed in empathy towards the world, the environment, the animals, and most importantly, the people who are all God's creations. My father was a good man, but he was a victim of his time, as were many people in history, and as we all will be when our lives become part of its course. We are cautioned not to judge people because by the judgment we make of others, so also will we be judged. We are to discern right from wrong and treat others with honesty, justice, and, most of all, love.

I have included my father's WWII diary that he wrote in Burma in 1944. The bracketed and parenthetical information is my insertion to clarify; otherwise, it is as he wrote it. A Diary Abbreviation Glossary is included to explain the common abbreviations he uses throughout, as well as additional reference charts to explicate matters of military life and rankings.

There are some insensitive references to our allies as well as our enemies that, if my father were alive today, he would profoundly apologize for. He was not a prejudiced person. When one is in the pandemonium of life and death on the battlefield, a person's blood and breath become the determiners of life. There is no greater empathy than to risk one's life for a brother-in-arms.

Diary Abbreviation Glossary

Ammo	ammunition
Ans	answer
ART	artillery
BAR	Browning Automatic Rifle
BCT	Blue Combat Team
Bn	battalion
CDD.	Certificate of Disability for Discharge
Cgt, Cgts	cigarette, cigarettes
Co	company
Cont	continue
Cp	camp
CT	combat team
Dpd	dropped
Erly	early
F	fever
Fr, f	from
Form	formation
GQ	General Quarters
75s	75mm Howitzers
KIA	Killed in Action
Mg	machine gun
Mt	mountain
MET	Medical evacuation tag
MIA	Missing in Action
Mishkina	Myitkyina
Mogong	Mogaung
Non-com	Enlisted military officer who was not commissioned
P&D	Pioneer & Demolition
Pl	platoon
Pos	position
Rat, Rats	ration, rations, food for mobile troops
C Rats	Canned, wet combat rations
J Rat	Jungle rations
K Rat	Daily combat food for individuals

10-1	Field rations parceled to provide 1 meal for 10 soldiers
RCT	Red Combat Team
Reld	relieved
Regt	Regiment
Rgt HQ	Regiment Headquarters
R	Rupee
Sqd	squad
Sta	station
Trans	transfusion
T S	tired and sore

Military Time to Civilian Time Conversion

Military	Civilian
0001	12:01 a.m.
0100	1:00 a.m.
0200	2:00 a.m.
0300	3:00 a.m.
0400	4:00 a.m.
0500	5:00 a.m.
0600	6:00 a.m.
0700	7:00 a.m.
0800	8:00 a.m.
0900	9:00 a.m.
1000	10:00 a.m.
1100	11:00 a.m.
1200	Noon
1300	1:00 p.m.
1400	2:00 p.m.
1500	3:00 p.m.
1600	4:00 p.m.
1700	5:00 p.m.
1800	6:00 p.m.
1900	7:00 p.m.
2000	8:00 p.m.
2100	9:00 p.m.
2200	10:00 p.m.
2300	11:00 p.m.
2400	12 Midnight

1944 WWII Army Military Ranks

Listed from Lowest to Highest

Private (Pvt)
Private First Class (Pfc)
(The rank of T/5, addressed as Tech Corporal, existed between 1942-1948)

Army Non-Commissioned Officers
Corporal (Cpl)
or
Technician Fifth Grade (T/5)
Sergeant (Sgt)
or
Technician Fourth Grade (T/4)
Staff Sergeant (S/Sgt)
or
Technical Sergeant Third grade (T/3)
Technical Sergeant
First Sergeant (1st Sgt)
Master Sergeant (M/Sgt)

Commissioned Officers
Second Lieutenant
First Lieutenant
Captain
Major
Lieutenant Colonel
Colonel
Brigadier General
Major General
Lieutenant General
General
General of the Army

Contents

Acknowledgments .... *vi*
Preface .... *viii*
Diary Abbreviation Glossary .... *x*
Military Time to Civilian Time Conversion .... *xii*
1944 WWII Army Military Ranks .... *xiii*
Contents .... *i*
The Interview 1992 .... *2*
Mary Evelyn 1943 .... *24*
India 1943 .... *42*
March to Staging Area 1944 .... *52*
Walawbum *71*
Nhpum Ga *102*
Myitkyina *131*
Mary Evelyn 1944 .... *155*
20th GENERAL HOSPITAL *158*
Linda S. Cunningham 2022 .... *169*
Epilogue *184*
Public Law 116–170— Oct. 17, 2020 134 Stat. 775 .... *190*
Merrill's Marauders Medal, from the United States Mint *195*
Diary of Larry W. Stephenson *200*
Works Cited *243*
About the Author *247*

The Interview 1992

Time: the ticks of a clock, the calendar days, is a human construct. It is not in God's wheelhouse, but His timing is perfect, I thought as I drove through the piney woods of east Texas, soon to be southwest Louisiana, the sun freckling my arm that rested on the open window. "El Condor Pasa" played its soulful strains on the radio, and I thought of the folklore of the condor that I had looked up years ago when Simon and Garfunkel had released it. From the recesses of my mind, I remembered that the condor was considered the King of Birds by Andean and Native American tribes and supposedly a good omen, but it was the meaning of the lyrics that I loved—the profound cry of the sadness of being bound to a place and unable to escape. Yes, my father was unable to escape. He had to fight for his very existence. He felt hunted by the Japanese and haunted by the vultures that waited patiently for the scent of death, both physically and spiritually. My father was more than the experience of war, although the war became the albatross around his neck.

I wanted to find out what made him tick. I wanted to know about the timing. I wanted to know about the tiger.

The wind blew in my face, clearing my thoughts so that I could be on top of my game when I interviewed my father about a war too big for words. He thought pain was a solitary thing, but it wasn't. We were all victims of that war. We were the 'home-front wounded' of a war that would never end. A war that changed just about everything in his life.

Once I crossed Highway 171, I began to see the familiar landscape of rice fields, green pastures dotted with cattle, and graceful, spreading oak trees. I thought of everything I knew about

WWII and, specifically, the Army's 5307th Composite Unit (Provisional) "Merrill's Marauders," a special-operations, long-range-penetration spearhead that went in behind Japanese lines in Burma, present-day Myanmar. The history books wrote about how the Japanese had captured Burma in 1942 and why Japanese Imperialism spread as dark as spilled ink on a map that engulfed the Philippines, Hong Kong, Singapore, Thailand, Indonesia, Malaysia, and others, but I was interested in the personal story from my father's diary that he kept meticulously every day but two of the campaign. The dappled light on the windshield mirrored my knowledge of it all. I was determined to find out today. It was my heritage, too. I drove cocooned in my thoughts, mixed emotions, and images of growing up with a dad who lived with a secret nightmare.

I remember as a kid, my sister and I climbed into the attic on a mission for anything that may work for a Halloween costume. Our orange tabby Bossy joined us. The flashlight made circles of light on the rafters as I crawled around the insulation, looking for the box that held the long, flowered skirts and puffy-sleeved blouses that would make a great gypsy costume. I bent over a beat-up, olive-green wood box and stooped down next to it.

My sister sneezed.

"I gotta get outta here; it's too dusty," she said as she glanced toward me. "That's Daddy's war stuff. You better not get into that. I got a spanking for looking in that box," she called as she disappeared down the ladder into the light.

I felt that I had been given an engraved invitation to look inside. The honor of your presence… After giving a two-second thought to the red stripes I would be wearing on my rear and legs, I was fumbling with the latch.

"Besides," I said out loud to Bossy, "I've had spankings before for dumb things, but this will be worth it. I'll just be careful to put everything back exactly as I found it. These are the secrets that Daddy never talked about. Secrets that make him yell and scream in his sleep."

The lid creaked as I lifted it and rested it against a rafter. I smelled stale, trapped air. A red tobacco tin with a picture of Prince Albert on it caught my attention. I opened the rusty lid, and a whiff of stale tobacco flared my nostrils. I carefully removed a black and white photograph of a Japanese soldier, a woman who was probably his wife, and a small boy. I stared at the family staring at me. The boy appeared to be a little older than my sister. The soldier was probably his daddy, and I bet that he was dead. His body, except for a missing finger that the Japs cut off, was cremated and sent to his next of kin. It rotted in a jungle in Burma or was picked clean by vultures. It occurred to me that my dad may have killed him. I gasped. My mind went wild over the revelation of my father's dark side, but it was survival. I could not stop staring at the woman with the creamy face and black hair and wondered about all the tears in such small eyes and how they must have flooded her smooth cheeks. My mama would have cried like that. I put the photo back in the tin and peered into the trunk once more.

I pulled out a long, curved knife. I didn't know back then that it was a *kukri or khukuri* in original Nepalese English. It must have belonged to a soldier in the allied Gurkha regiment that Dad had traded something in exchange for. I found a bayonet and wondered if it had protected him from a banzai attack. I wondered if he had ever held it over a campfire to roast game. I found a canteen that carried his water. Wrapped in a Japanese flag was a tan-colored hand grenade, presumably Japanese, and an American olive-green hand grenade. The tops were unscrewed, and there was nothing in them. I sat for a moment and thought about rocks as weapons. I remembered my preschool days when I got in trouble for throwing rocks at the school bus. I thought of cavemen throwing rocks and stones for survival and how it evolved into the hand grenade. I took the flag out carefully because it was torn. It had a big red circle on it and was covered with a dark brown stain. I quickly surmised that it was blood, not just American or Japanese, but human blood. I wondered who used a flag to stop blood from flowing. Whose blood? I guessed that it was my dad's. I wondered about my own blood and the arteries and veins that kept it contained. Under a stack of letters addressed to my mother in my dad's handwriting, I found a small, black diary. I gasped at the thought of my questions being answered and about all the things Daddy wouldn't talk about. I opened it carefully and read dates, miles marched, and more, but the print was too small, even the flashlight wasn't enough for me to decipher it. Instinctively, I knew I had found a great treasure. At that time, I knew nothing of the tiger that paced and snarled.

I turned down a side road and headed for the main highway. As I passed a neighbor's house, I realized that Sachiko was outside picking vegetables from her garden. She looked like a Monet painting with her straw hat and basket, surrounded by colorful vegetables, but the plastic grocery bag was a dead giveaway of modernity. I stopped, backed up, and pulled into the driveway. Sachiko straightened her back and looked up, saw me, and waved. I walked down the row boasting its abundance of red tomatoes, yellow and green peppers and met her halfway.

"Konnichiwa," I said, "How are you doing?"

She adjusted her straw hat so she could see me and smiled at my attempt to speak Japanese. We bowed to each other in the Japanese tradition, and she patted me on the cheek.

"Good. Long time since I see you. You, okay?" I nodded.

"How is Mr. Rich?" I asked.

"Good. Always busy."

"How are all your kids? Gone by now, right?"

"Oldest married. Hank joined Army, then Tommy and Hana. Youngest boys work, and my baby girl is now married."

"So, Hank, Tommy, and Hana joined the military?"

"Tommy always does everything Hank does. Rich was in Army, and your daddy was too. I guess that's why."

"Well, good for them! All your kids are grown now. It's hard to believe."

"Old Japanese saying, 'Even when months and days are long, life is short.'"

"That is so true. When you talk with them, tell them that I said 'hello.' I remember Hank and Tommy the most. I remember the fun we had when we were kids, riding our horses, and that little burro, Danielle, and getting bucked off."

"I remember. You all wild kids," she laughed.

"Mr. Larry and Ms. Evelyn, okay?"

"I think so. I'm on my way there now to check on them."

"I go check on them every other day. Mr. Larry, he mellow now. When he first moved here long time ago, he did not like me."

Her last sentence ended with a thud, and I didn't know what to say.

"Take to them," she said, and she offered a plastic bag of tomatoes. "I will," I said. "Thank you. Good to see you."

She nodded and patted her heart. I walked back to the car in puzzled thought.

I thought about Daddy not liking Sachiko. It had never occurred to me before. He comes home from the war, goes to work, builds a house in the country, and his neighbor is Japanese, the very race of the people he was trying to get away from, trying to forget. She was probably a constant reminder of the war. I continued down the road and turned onto the highway. The mailbox with Stephenson and the number printed on it was just ahead. Their three-storied home with the French colonial architecture that Daddy had built himself came into view, a testament to his larger-than-life personality and his ability to think big. The sun shone brightly as I pulled off the road into the subtlety of shadows and drove down the long driveway, admiring the daylilies that deserved a slower, walk-by look. The daylilies trumpeted their butter-colored blossoms upon my arrival, and the spreading live oak branches scooped my imagination into its arms and welcomed me home. I hoped my interview would be fruitful and healing. After all these years, Daddy decided to talk about it.

"You are the only one interested," he had said. "I have

encouraged you for so long in your writing; now you can write the heck out of this story."

My father was a war hero, but he didn't know it and wouldn't acknowledge it.

"I was just doing my job," he once told me when I broached the subject with him.

Mama came to the backdoor wearing a denim skirt and white blouse; her make-up was understated. It was a far cry from the shirtwaist dress, pearl earrings, lipstick, and hair perfectly coifed, a throwback from the *Leave it to Beaver,* 1950s mom. She held the door open as I gathered my recorder, a few maps, and books and exited the car.

"Hi, Mama," I said as I kissed her cheek.

"Hi, sugar. How was the drive?"

"Long, but beautiful as always. Is Daddy okay?"

"Sure, he's fine. Looking forward to visiting with you."

"Great. Sachiko sends you tomatoes," I said as I handed her the plastic bag. She took it and my rolling overnight bag, and I juggled the rest.

Daddy sat in his brown, nubby easy chair. The plaid shirt I had given him last year for his birthday draped on him like a flag on a windless day. A red line of the scar from his heart surgery snarled on his white, wilted chest just above the second button of his shirt, but his blue eyes didn't just twinkle; they snapped, crackled, and popped. Yes, that was my red, white, and blue Daddy.

I breezed in, swooping on my dad with a big kiss on the cheek.

"Hi, Dad, how are you doing?"

"As good as gold, honey girl. Can't complain."

"I see that this rocker is still squeaking. I'll get some oil for you."

"No, don't do that. As long as I hear the squeak of the rocker, I know that I am still alive," he said with a wink and twinkle in his eyes.

"Okay, squeak away."

Dad had a chess game that stayed set up with the last move of the game that we had started and hadn't finished since the last time I was there.

"It's your turn," he said.

I glanced at the chess table. The light from the tall windows streamed in and brightened the room. His trophy mount of a huge elk stag that he called 'Willie Gus' hung on the wall, a testament to his younger hunting days and his older changing ways.

A rush of memories filled my mind while I pondered my next move. I remembered starting out on the train trip that we took years ago. I was viewing the memory as a ten-year-old with an adult perspective, even though I perfectly remembered it.

War is a curious thing, I had thought as I watched soldiers gather their gear and mount the steps of the Southern Pacific Railroad Station. If people didn't die, the soldiering business would be kaput. Maybe that's why there are no wars in heaven, I mused to myself. Soldiers were on their way home. There was troop movement everywhere even though WWII ended in 1945 and the Korean War had died down in 1953. For some of the soldiers, there was an anger present that stayed on simmer for the most part. I could see it in their eyes. Others were worn out. Their eyes stared forward, the thousand-mile stare, but they really looked backward into their brains that were muddled with landmines of memories. I remember Daddy used to look that way—still did sometimes, but not as often. Other soldiers gawked at the ladies as if they hadn't seen one in a terribly long time. They whistled, smiled, bowed, and offered to carry luggage, telling the ladies they were the most beautiful dames they had ever seen.

They said, "Show me to the altar, baby. You are the woman of my dreams," and "How about some backseat bingo?"

I didn't understand their jiving talk, but I watched and listened to them so that I could understand my daddy. I supposed the soldiers of WWII acted the same way as these soldiers. I exhaled a sigh and plopped down on a bench next to my sister, who sat next to my little sister, who sat next to my little brother, who sat next to my baby brother, who sat on Mama's lap. We all waited for the train while Daddy went to the ticket office. Fat raindrops began tap dancing on the roof. Typical weather for Southwest Louisiana. I swung my legs, admiring my shiny black patent leather shoes and my white lace socks. We were going to Arizona, where my daddy's mama and papa lived. School was out, and Daddy had saved his pocket change every day for a year in a bank that he made from an old breadbox with a chute for the coins that he placed under the breezeway steps. I remembered the time my sister and I pulled the concrete steps away from the wall and took a nickel each for the ice cream man. Then we couldn't push the steps back, so we called Grandpa to come over and push it back before Daddy got home.

Grandpa did and never told Mama or Daddy. What a close call. I clipped that thought next to the rock-throwing incident in my brain to remind myself never to do that again. I also remember one evening hearing Mama whisper to her sister that 'it would be good for him to visit and get a change of scenery.' I surmised that she was talking about Daddy.

My attention turned, and I watched a thin, happy-or-drunk soldier finish his can of beer with his buddies and smash it on his head.

"That's the only thing your head is good for, Pete," joked his soldier-buddy with the curly hair.

"Naw, man, the bullets just bounced right off of it," Pete said. "Mama used to call me 'ironhead.'"

"She called you 'ironhead' because you were too stubborn or dumb as a block of iron," Curly Hair said.

"You think I'm dumb? Well, watch me with the ladies, son!" Pete said. Pete stood up and did a dance around, then saw us sitting like little birds on a branch.

"Look at the little ankle-biters all in a row," Pete said as he walked in front of us. "And look at the beautiful mama," Pete continued and gave a cat whistle. Mama turned her head and ignored him.

"We ain't ankle-biters. I wouldn't bite your dirty, old ankle," I said and stood up with my hands on my waist.

"Don't talk to him, and don't say ain't," Mama said. Pete turned in surprise and looked at me.

"Look at this little whippersnapper," he said to his buddies and laughed a big laugh.

"I am not a whippersnapper. My name is Maggie Anne," I said indignantly, tossing my hair.

"Raggedy Anne," Pete said, misunderstanding my name. "Named after a rag doll."

Mama stood up and waved to Daddy, who was standing with the luggage, waiting to get on the train. She handed my baby brother to my older sister.

"Kids, follow your sister to Daddy."

We left in line with Mama following me so that I couldn't backtalk that old, drunk soldier. Mama held her closed, long umbrella under her right arm. Her hand held the handle. I could see in the reflection of the long ticket office window that we marched in a column, but curiously enough, Pete followed Mama. He walked, leaning backward with his knees bent and wide open. His hips were pushed forward and gyrated right near Mama's behind! Mama must have seen the reflection because she stopped suddenly and jammed the umbrella backward into Pete's crotch. He fell with a loud yelp and doubled over. His buddies started whooping and laughing their heads off.

"Some ladies' man," Curly Hair said, and the soldiers began laughing again.

Mama hurried us on our way.

"You little firecracker." Daddy laughed and whispered to her when we met up with him.

"I learned to take care of myself when you were off at war," she said. "Remember when I quit my job at City Service because of a similar incident?"

We settled in our seats, and I wondered why I had not openly spoken to that soldier. As an adult, I understood. As a child, my voice was muted by the mores of a society that did not value a child's perspective. That entire back-and-forth exchange was what I wanted to say to him but did not. It was all in my mind because Daddy taught us not to fight or be confrontational, and Mama taught me and my sisters to be ladies. Daddy said on occasion, "Fighting leads to hatred. Hatred leads to war. That is why our ancestors encouraged civility." Also, my name is not Margaret Anne or Maggie Anne, or Raggedy Ann. My name is Linda. When I was young, I was precocious only with familiar people.

I heard the train whistle blow, and we heaved forward, gradually moving through the railroad yard and picking up speed. Once we were on the outskirts of town, the train was eating the rails at a stiff clip.

"Windy Lindy," Daddy said, "Let me show you some more chess moves." I jumped up and followed Daddy to the empty observation car. We sat down at a table, and he pulled out the chess board and a zippered bag with the chess pieces. He set the board up, and we began with our opening moves. "Chess is kind of like war," Daddy said. "You gotta see the big picture, but you should pay attention to the individual details. Look for the weakest spot. You can see the whole board, right?"

I nodded my head.

"What if you were in a jungle and there was bamboo that blocked your vision?"

He took a bud vase with a flower in it and put it in front of my knight. "What if you couldn't see around it? What would you do?"

"Well, I guess," I said as I chewed my lip, "I would watch what I could see like your Queen."

"Good," he said. "If you can see all of the other pieces, which piece is behind the vase?"

I looked at all the pieces, counting his and mine in my head, and said, "Your bishop."

"Good. You must learn to think. In a jungle, you learn to listen to birds or monkeys chattering and creating a racket. Something is disturbing them, maybe the enemy soldiers. You have to use all of your senses, even the sixth sense, your intuition."

"There's nothing to listen to in chess, Daddy," I said.

"Yes, there is. Listen to the noise your opponent makes. Is he shifting in his seat? Is he sighing? Is he sitting extremely still? Is he holding his breath? Is he rubbing his hands together?"

This memory flashed back through the years in a few seconds, and here I was in the present once again, talking with my old dad.

"Dad, do you remember on that train trip to Arizona when you taught me to play chess?"

"I do remember that. You picked it up so fast you beat me," he said. "Awe, you let me," I smiled. "But I didn't know that then."

I dug into my purse. Daddy looked at me, amused. "The Black Hole of Calcutta?"

"Yes, actually, my portable trash can," I laughed.

"I read where the Black Hole of Calcutta, India was an actual hole in the ground where they put British prisoners of war and Indians who were sympathetic to the British after the fall of Fort William sometime in the 1750s. It was so crowded that many died of suffocation and heat."[1]

"Well, my purse is just as infamous. Sometimes I think something's going to bite me if I stick my hand in there," I joked. "Here it is," I said as I pulled out a chess piece—a knight with a red cross painted on his breastplate armor that I hoped would jog his memory, but my feeling was that his mind was sharp as a kukri knife.

"Do you remember this?"

"I certainly do. I gave that to you on the train all those years ago," he said. "I found that when I was in Deolali, India. One of the British soldiers must have dropped it. I kept it in my pocket for the entire campaign."

"I found it the other day and thought you might want it back. What's the history of it?"

"That's Galahad, the one that searched for the holy grail. Our 5307th unit 'Merrill's Marauders' code name was 'Galahad.'"

"Why the name Galahad?"

"Sir Galahad was the illegitimate son of Sir Lancelot and Elaine of Corbenic. That's why I think Michael Gabbett named his book about Merrill's Marauders, *The Bastards of Burma.* Gabbett also claimed that, like the bastard children of English and French royalty in the Middle Ages, the Marauders had to fight for recognition. Galahad was known for his goodness because he was an orphan and grew up in a convent of nuns. He was one of the three knights that found the Holy Grail. The knights were formidable, mounted warriors who were organized as a shock attack group. The Germans also had shock troops trained for mobility and could penetrate an overwhelming enemy from behind enemy lines using the element of surprise. They were the tip of the spearhead. High casualty rates were expected."[2]

"Apropos," I said.

"Right. Someone knew ahead of time that we would have a high casualty rate. I was just a pawn and expendable at that. We were all expendable."

"What does it mean to be expendable? I asked.

"It means that you could be sacrificed in order to accomplish the mission."

"Like a suicide mission. I read where you were sometimes called the magnificent 'misfits' of a secret jungle mission, but I have never considered you a misfit."

"No, I was not one of the misfits. I spent six years, eight months, and twenty-three days in the military, including the time in Merrill's Marauders. We got that name because some of the guys volunteered from military stockades on Guadalcanal and some of the other islands. These guys weren't your 'spit and shine,' 'yes sir, no sir!' type of soldiers. They were unconventional, but they were hell-raising guerrilla fighters."

"Dad, I'm going to record you because what you went through was historical and so heroic that it was almost mythical. If at any time you feel like we are getting too close for comfort, I want you to stop. I don't want you to freak out. I'm doing this for posterity and your legacy as well as Merrill's Marauders." I pressed record.

"I would prefer for the legacy of Merrill's Marauders," he said. I nodded, and Dad began.

"I'm not sure whose brainchild this was, but I do know that it was a coalition effort between Prime Minister Winston Churchill, Lord Louis Mountbatten—who was head of the Southeast Asia Command—and President Roosevelt with his Chief of Staff General George Marshall. Roosevelt and Churchill met at the Quebec Conference and again at the Arcadia Conference in Washington, D. C. From this meeting, they asked Generalissimo (G Mo) Chiang Kai-shek to be the Supreme Command of the United Nation's China Theater.[3]

"There was a tug of war between the British Major General Orde Wingate, who was the first to organize the deep penetration into Japanese-held territory in 1942 with the 77 Brigade, which was later named the Chindits, and the American General Joseph Stillwell, also known as 'Vinegar Joe,' who commanded two division-sized units of Chinese fighting in Burma in 1942.[4] The Chinese divisions were no match for the powerful Japanese aggressors and General Stilwell had no option but to retreat. He and his staff of 117 men walked 140 miles to Assam, India, because of the 'hell of a beating' they took from the Japanese.[5] This calamitous loss for General Stilwell's first command made him want to redeem himself. I think General Stillwell was chosen because he had the fire in his gut to go back and win, even though he was 59 years old and blind in one eye. He was a West Point graduate, and he spoke Chinese. General Frank Merrill was also a West Point grad and was probably chosen because he spoke Japanese and he had the reputation of being a soldier's general because of his affable personality to connect with his men. I can tell you we would have run through a brick wall for General

Merrill—as it was, we overcame the impossible, which was more difficult than a brick wall.

"We were organized into three battalions of 1,000 soldiers each. We were further divided into combat teams. In the beginning, we were called the A, B, and C shipments because it was a secret mission, and that name indicated supplies instead of troops. Once we were in India, we were called the Alpha, Bravo, and Charlie Battalions but were later renamed the First, Second, and Third Battalions. The battalions were further divided into two combat teams. First Battalion had the Red and White Combat Teams. The Second Battalion had the Blue and Green Combat Teams. The Third Battalion had the Orange and Khaki Combat Teams. I was in the Third Battalion, Orange Combat Team, Company L."

"I didn't know this back then, but the military prepared for almost everything except how they were going to get us out. They didn't even have a plan for the evacuation of the wounded until we were there and prepared to engage the enemy. There were 2,997 total soldiers; however, 247 remained in India for operations. The top brass expected an 85% casualty rate which would have been 2,337 men out of 2,750, leaving 412 to survive, but we had a casualty rate close to 80%.[6] At any rate, between 500 and 600 exhausted and sick men survived, and out of those, only 130 were able to fight when they took the airstrip at Myitkyina; however, 200 additional men who were convalescing in hospitals in Ledo were ordered back to the battlefield for the final mission to take the town of Myitkyina. There were more casualties from disease than from death. Record-keeping was discouraged by General Merrill. I don't know how many men who were sent back to the States succumbed to their diseases and wounds."

Dad paused and stared out of the window as if he had made a

giant leap in time.

"I have often wondered why and how I survived, and so many didn't. I thought then, and I still think, 'What am I supposed to be doing to deserve the gift of life?'"

"Maybe you are doing it now," I whispered and swallowed the lump in my throat.

I felt we might be getting too close for comfort—mine or his, I wasn't sure—so I changed the subject.

"Dad, can you tell me about your earlier years and when you decided to go into the military?" I asked.

He stretched his legs out and thought for a moment.

"I was a paper boy and read the news as I folded papers. Early every morning, I sat outside under the old cottonwood tree. It provided enough shade to make folding newspapers in the blistering Phoenix sun tolerable. The chitter chatter of the leaves and the moving patches of sunshine made the world that surrounded me feel magical—the news of the day made the world smaller, even if we were in hard economic times. I didn't realize we were poor because everyone was in the same stinking, sinking boat of the Great Depression.

"I was full of spirit and spitfire, ambition, adventure, determination, and duty, even though it was 1931. We couldn't do much back then but dreaming up a grand future was a great pastime. Mama had the radio on, and the lilting music of "As Time Goes By" from the Casablanca movie floated on the breeze. It made me think of time passing and what I wanted to do with my life. I thought about when I was born and how I grew to be a pipsqueak and then a teenager. I was thinking about what I wanted.

I decided I was determined to do something important and never want for money. I knew what I wanted to do but did not know why. I figured God would let me know when He wanted me to know it."

"I fell into a reverie as the lyrics interjected thoughts of love. That dreamy music made me think of girls. Oh, how I wanted to fall in love with a beautiful girl." He looked at Mama and winked. "All I wanted was a new set of clothes. I tried to save up my money from the paper route, but so often had to share with my mama when ends didn't meet. That followed me all the way to high school, where I joined the Junior ROTC just for the uniform. Why? To impress girls."

Mama smiled and gave a little amused laugh.

"One early morning, before school started, my fellow ROTC guys and I marched out to the flagpole at 7 A.M. In ceremonial sincerity, I unfolded the flag, attached it to the flagpole, and raised the red, white, and blue. The flag snapped in a gust of dusty wind. Smiles Smith—the bugle player with the big grin, except when he played "Reveille"—began the morning ritual. He had always struggled with the rhythm and tempo of "Reveille," but this morning, it was perfect. The dust settled, and the flag looked crisp and bright in the morning sun. I had goosebumps on my arms and up my spine. I felt a surge of pride for The United States of America and for the United States Military. I knew then that I was going to be a military man."

"When I graduated from high school at seventeen, I drove a Holsum Bread truck for a few months until I got impatient, and my friend Charlie and I joined the Army on September 13, 1937. That same day I broke the news to Mama. She stood at the kitchen sink, looking out of the window while she washed dishes. A pot of pinto beans on the stove chuckled and gurgled when I came in full of bravado. I pretended to bugle "Reveille" and stopped when Mama turned around. "I have an announcement!" I declared.

Mama wiped her hands on her apron and turned to the stove to lower the fire.

"Well, whatever it is, it sounds important," Mama said.

"Your son is now a military man," I said.

"Charlie and I joined the Army."

Mama had no words for a few seconds. She stood in her diminutive self and drew a deep breath.

"I'm not surprised. Frankie joined the Cavalry, and Buddy joined the Marines. I suppose Oscar will be following in your footsteps. But aren't you too young?" she fretted.

"Oh, it was this matter of a slip of the tongue," I said.

"You lied about your age?" Mama questioned.

"Something like that."

"Will, you are only seventeen. Your Papa is not going to like this."

"I know, Mama, but I've got it in me. It's my destiny."

"I know, son, but I don't have a good feeling about this. The world is a worrisome place."

She picked up her spoon and stirred the pot of beans, and continued, "But what's done is done. You have always been a good boy, confident, and responsible. Continue to make responsible decisions and stay true to yourself. That's a hero's job, and remember what I've always told you, 'Mind over matter.'"

"A week later, we took off in a bus that was filled with new recruits, Mama's words ringing in my ears. We were on our way to boot camp, where I did everything they told me to do. I canvassed the area for cigarette butts when commanded to do so. I put great effort into my uniform and made a quarter bounce off my tightly made bed. I ran hard and fast for short distances. And then slowed to a jog for long distances. We rappelled from obstacles and climbed over and under them. We belly-crawled under live fire. On the combat firing range, I made expert. I had been shooting rattlesnakes since I was a kid, and I was a good shot. One of the most important things that I did was to show the drill sergeant my ultimate respect. At the Pass and Review ceremony, our class marched on that field with a smart step and perfect timing. With our boots and brass shining, we stood straight as an arrow when we saluted the General. My heart felt like a beating drum roll. I was so proud to be part of the legacy of those who gave their lives and fought for our nation to be free. You see, life is divided into two groups of people: the takers and the givers. I was determined to be one of the givers because Mama and Papa raised me to give the shirt off my back."

"Daddy, why did your mama call you Will when everyone else called you W.L. for Winford Laurie?" I asked.

"There was a crazy old man named Willie Gus that lived down the street. Mama tried to befriend him through kindness, but she realized that there was nothing she could do to help him. She told us to stay away from him, so she called me Willie Gus when my antics were driving her crazy. As I grew up, I would do stupid things that would get a rise out of her. It became funny banter between us. So, Willy Gus was shortened to Will. My mother was a reader, and she liked the neighbor boy Laurie in Louisa May Alcott's book *Little Women*. So, she named me Winford Laurie after him. I changed my name to Larry Winford Stephenson because I considered Laurie a girl's name, and the kids teased me about it."

"Kinda like Johnny Cash's song, 'A Boy Named Sue?'"

"Yeah," he laughed.

"We were tough guys back then."

Mary Evelyn 1943

"Dad," I said, changing the subject, "I have always loved love stories.

Growing up, you and Mom gave us the example of love. Can you tell me about falling in love with her and how it helped you to survive?"

He rocked in his easy chair a moment and stopped.

"Falling in love is so hard to describe. Hmm," he mused. "Falling," he began, "I would think it means to be helpless, overcome, or accidental. I can think of the cartoons with hearts exploding and enlarging out of the chest of a mouse, Mickey, that is."

He looked at me with his blue eyes sparkling with humor.

"It's funny because you feel so dumbfounded, so silly, and stupid, and self-conscious. That's one kind of love. Real love is different. It felt bigger than me, than her, or both of us together. It felt like a fog that surrounded me, and she was the light. She was coming toward me as if in a dream that I dreamed long ago. I felt fate was giving me a blessing to pursue this girl. It was God-given and would penetrate time. Falling in love was like falling into the divine, into heaven, into the arms of love. I had found my home when I met her, but it was more than that. I found my future; I knew that if I had to die, I wanted to die in her arms."

"I was assigned to Company K 158th Infantry 45th Division Training Center at Fort Sill, Oklahoma. Fort Sill is located on a plateau near Medicine Creek, which is in the foothills of the rugged Wichita Mountains known to be the Apache Chief Geronimo's hideout. Geronimo's gravesite is in the Apache Cemetery there."

"Dad," I said. "Pat was stationed at Camp Eagle in the Wichita Mountains at Fort Sill when he was in ROTC in college."

"What a coincidence!" We said at the same time and laughed.

"I never knew that I would have a son-in-law train there years later," Dad chuckled.

"Continue with your story and how you hooked up with Mom," I prodded.

"Okay," he began, pausing to collect his thoughts, "We slept in tents at night and trained for combat in the great outdoors by day. The fall at Fort Sill was beautiful, blustery, and a little chilly in the mornings. The clouds were fluffs of bright cotton candy in the sky—sometimes, they were white, and other times they were pink, depending on the time of day. The moon at night? Well, it was a lover's moon when it was full, and I was terribly lonely. I was a homesick kid—so young back then. Too young to be in the Army. I rarely received any mail because Mama and Papa were in California working a construction job and living in tents as well. My sisters were busy with their families."

"The best stroke of luck happened when the 158th Infantry of the 45th Division participated in the Louisiana War Maneuvers before I went to the Panama Canal Zone from January 1942 until January 1943. When we were in Louisiana, I was stationed at Camp Polk—today's Fort Polk."

"I met Evelyn in Elizabeth, Louisiana, where she grew up. One weekend, I was on MP duty for a group of soldiers that had a pass, and we boarded a dilapidated school bus at Camp Polk. That bus sounded like a machine gun as it rata-tat-tatted us all the way to

Elizabeth. We had heard that a Bulldog baseball game was that afternoon, and the star pitcher had been written up in the news as the next Spud Chandler of New York Yankee fame. I was there, you know, trying to keep the men out of trouble. Allen Parish was dry, but having grown up with a father who was a bootlegger in the 1930s, I was suspicious. When I was little, my Pop rigged pipes that came down inside the wall the length of the stairway and dripped into a jar in a secret compartment. I stuck my tongue under the pipe one time, and I'm here to tell you; it would curl your hair. Pop called it the 'upstairs articles.' We were sworn to secrecy. I'm not making excuses, but during the Great Depression, it was the only way he could provide for a wife and seven little pranksters. The revenuers finally discovered Papa's secret whiskey still and shut it down. It was for the best because Papa was becoming his own best customer. He later sobered up with Mama's help. Anyway, we drove into Elizabeth, which was a quaint town with its neat, white houses with front porches. We called it the front porch town."

He sat there quietly for a few minutes. He was there; I could tell. Time had slipped through his fingers, and he was that dashing soldier with a beautiful smile and bright, blue eyes.

"Stay outta trouble, boys," I said as the soldiers exited the olive-drab green school bus. "Be back after the ball game, or you will be hoofing it back to camp."

"Yes sir," they said and took off.

"J.D., let's go check out the town for places where the guys could get into trouble," I said.

"You bet. Wish we could meet some girls," J.D. said. "The odds look a little slim. It's a small town."

"Too bad. We are all spiffed up, and no girls to appreciate it," J.D. laughed.

The sun was hot except in the shade of a tree-lined dusty road that we walked on. A few cars were parked at an angle in front of storefronts that appeared to be the town center. I saw a Coca-Cola sign at Finke's Department store.

"Let's get a Coke," I said. "It's not exactly bottled sunshine, but we can't drink on duty."

We strolled in and paused to look around for the soda fountain counter. I took my sunglasses off and leisurely turned. There, the most beautiful girl I had ever seen swung around, almost in slow motion, her dark hair catching up with her head, and she paused when she saw me. She wore a yellow dress and looked like a delicate canary perched on the stool. A friend sat next to her, who later turned out to be her sister, Ethel.

J.D. whispered, "I've got the one on the right."

I smiled, nodded, and casually walked to the fountain.

"Is it always this hot in the piney woods?" I said, looking at the beauty on the left.

"In the summer—yes, but I promise you, it can get hotter," she said.

"Her name was Evelyn, Mary Evelyn. I didn't know it then, but in Hebrew, Evelyn means 'life.' She was my lifeline. We struck up a conversation, and I asked if I could write to her. We wrote for two years before we married. Her letters and prayers kept me alive."

"Mama, what is your take on this?" I asked.

"After we met, Larry came to see me when he was on maneuvers in Louisiana. When he had a pass, he came by and met my family. My daddy worked at the paper mill in Elizabeth and took him on a tour of the mill. But for the most part, we got to know each other through our letters. From Fort Sill, he was sent to Camp Barkley in Abilene, Texas, for further training. I remember receiving a few significant letters that I kept and saved for occasions just like this. I recognized that they were historical," Mama said as she fumbled through a box filled with a stack of letters.

"Here they are," she said and handed the letters to me. I read them out loud.

November 18, 1940

Darling Evelyn,

Gosh honey, I was glad to get your letter. I'll bet I have read it through at least ten times. It makes me feel so good inside to know that you do love me.

I have a small portable radio turned on now and the music makes me so lonesome. The song playing right now is by Benny Goodman — "Is that the Way to Treat A Sweetheart." I wish we could go dancing and have fun. We are young and alive. We don't know what we will face from one minute to the next, but together we can face anything. Oh darling, I hope to get to see you soon.

I don't know just when I will be able to get my pass, but I will send a telegram just before I leave for Natchitoches and for your sake, I hope it is before Christmas. I might have to wait until New Years.

Darling, Saturday I had the highest honor I believe I will ever get. The General Brees of the Third Army appointed myself and five other sergeants of K Company to investigate the Lawton Oklahoma police force and to find out if they were dishonest. I turned in my report this morning to the General Brees and it was favorable. I'll close now. I love you so much darling and I'll never stop loving you.

All My Love,
Larry

P.S. Keep your chin up and the sparkle in your eyes.

"That is so sad to be in love and to never see each other or hear each other except to communicate by letters," I said.

"Back then, that's the way it was. We had nothing to compare with it. I guess we had a little more patience than people of today," Mama said.

February 19, 1941

Dearest Evelyn,

Gosh Darling, I miss you. I certainly wish you were here with me now. Darling, I have been transferred to the Government Secret Service. It was certainly an honor to be picked out of the 45th Division. It is a dangerous job. You don't know from one minute to the next what is going to happen.

Darling, I am not supposed to disclose my destination or where I am and most important, I am not supposed to write anyone or receive any mail for it would be dangerous for me if any foreign agents were to get a hold of it. Please honey, it would be best to burn this letter when you get through reading it.

I love you with all of my heart,
Larry

"I was in the dumps," Mama said, "I remember walking to get the mail at the end of the driveway and thinking that the weeping willow tree hung over with an invisible weight just like I did. I felt the weight in the slump of my shoulders. My heart felt the weight of another empty mailbox. It had been three months since Larry's last letter. I couldn't write to him, nor could he write to me. I had no idea where he was or what he was doing. I refused to let my imagination run away with me. In my mind, I saw Larry laughing and smiling. He was the happiest person I knew. That's the way I thought of him, but oh, how love hurts. Then the day came, and I opened the mailbox and saw Larry's letter, and pitiful me perked right up."

May 1941

Dearest Evelyn,

Darling Evelyn, I was glad to hear from you. Please don't ever worry about me because honey I'll never let you down as long as I live. I can't get a pass for a while yet. Maybe two months but as soon as I get it, I will see you and I hope to God it's soon.

Well honey, to tell you more about the work I was doing, if you have ever heard about the Army Secret Service it's the same as the F.B.I. It is the Army Secret Intelligence Branch known to us as G-2.

You will know there was quite a bit of sabotage going on at Douglas Aircraft in Los Angeles, and I and three other agents were trying to stop it. Well, we did catch two of the spies, Vogel and Harbin, two Germans, and then, they found out I was just 19 and transferred me back to the company as being too young. Since I have been back to the company, the guys have been treating me like some kind of God. They do anything for me. It's really swell.

Please Darling give my regards to your folks and tell them I'll see you and them as soon as possible. I know one thing; we are maneuvering down there this fall.

That's really great because I'll get to see you quite a bit then.

Oh honey, I'm glad you are going on with your career. You have what it takes to finish. I'll close now with all my love. XXOO

I love you,
Larry

P.S. Please answer as soon as possible and one thing I forgot to mention, I just got promoted to sergeant because of the excellent work I did in Los Angeles.

"Wow, Dad. You were in the Secret Service?"

"Just for a short while. Back then, you could join the military when you were twenty with your parent's permission. The draft age was lowered in 1942 to eighteen."

December 7, 1941

Darling Evelyn,

I hope you are over your mad spell tonight for as I have said a million times, it isn't good for you or in fact, anyone. Well honey something happened today. I saw the show <u>Sgt. York,</u> and when I came out at about five-thirty, I heard that Pearl Harbor and Manila Bay had been bombed by Jap planes. The U.S. is supposed to declare war in the morning. I don't know honey, but I have been drilled for a long time and I want action now. Remember darling if anything happens, I will always love you and some time or another, I will come back to you.

Another thing darling, no more passes or leaves are available until further notice. All men on passes and furloughs must report back to their bases immediately.

Give everyone my best regards.

I love you, darling,
Larry

December 10, 1941

Darling Evelyn,

What do you think of the war now? Well, don't worry about me Sweet, for they probably won't ever send us over. For the past two days since war has been declared by the President, we have intensified our training from firing on combat ranges to combat target—everything that goes with modern warfare. Tonight, I am going to listen to the president's speech. Isn't freedom wonderful darling? Just think, freedom of speech and everything that goes with it. I would gladly fight for that. Most of all I will fiercely protect you and our love for each other. Love rouses the fierceness of the tiger in me yet love makes the lamb lie down beside it.

I'll have to close Darling for if I write a book, I'll have to have it bound. I love you Darling and I wish you were in my arms tonight.

*Love,
Larry*

"Yes," I agreed, "these letters are a treasure trove worth saving. It's funny when something momentous happens, your brain snaps that moment in time as if it were a photograph. I remember when JFK was killed. I was in Spanish class when the principal made the announcement over the PA system. We sat stunned, and my friend Lola began crying. The reality of our President's death made my heart wrench."

I handed the letters back to Mama.

"After the bombing of Pearl Harbor, I thought for sure Larry would be sent overseas to fight, so we decided to get married during his leave," Mama said. "We married when your Daddy came home from Panama. He arrived home on January 26, 1943, and we were married on February 24, 1943. We moved to Clinton, Louisiana, for several months because your dad was stationed in Mississippi. It was in September that he volunteered for the special, secret mission. They told him he would be out of the Army in ninety days. Of course, we did not know where he was going because it was a secret mission. He left San Francisco on September 20, 1943.

"I miscarried early in September, right before Larry left. I was shattered after I lost the baby. We both cried because we wanted a little one, and I especially wanted a part of my husband when he went off to a war from which he may or may not return. My Daddy, your grandpa, thought it would be best for me to stay with my Granny in Dry Creek so that I could heal emotionally and physically. Daddy drove me because, like everyone, we only had one car. I remember when he turned down the dusty lane to his folk's home in Dry Creek, Louisiana, and the old homestead came into view. The dogtrot-style home set among tall, spreading oak trees loomed in the distance. I loved that old place and thought of it as a brown topaz jewel surrounded by bright emeralds.

"This was the Hanchey family home. Dry Creek was the town of my birth, and the homestead was the place of my dad's birth. I was born a few miles away in the Kent family home that burned to the ground many years ago.

"Dry Creek, which was south of Deridder, Louisiana, was a rural farming community where my parents grew up, and my grandparents farmed. When I was young, Daddy got a job with the paper mill in Elizabeth, Louisiana, where we moved and where my siblings and I grew up and went to school. I missed Granny and our talks and how she simplified the world of chaos. The simple wisdom of a woman of faith was an undisputed reckoning. Granny had taught me the names of flowers, how to walk in the woods, and where to find Indian artifacts. She always saved the cream from the cow's milk just for me, her firstborn grandchild. Granny came to the front porch wiping her hands on her apron. Daddy stopped the car gently and waited for the dust to settle, and I savored the sight of my old Granny on the porch anticipating my arrival. The long porch was comforting because it was about inclusion; it was about family, and the tales I heard repeated over and over, like the hole in the floor where grandpa, your great grandpa, was cleaning a gun, and a round went off by accident. That hole became a reminder to always double-check the chamber before cleaning."

Daddy greeted his mother and went to look for his Papa. I hustled up the steps and gave Granny a big hug.

"Oh, how I have missed you," Granny said. "I've missed you, too."

We walked toward the porch swing when I noticed the purple flowers that spread their tendrils on the post of the porch.

"Granny, that vine with the purple flowers is gorgeous. What is it?" I asked.

"Why, it's a passionflower," she said to my fond amusement of using 'why' as an interjection. "I planted it a few years ago, and it has really taken off. It symbolizes the story of Christ and is a wonderful way to teach His passion to children so that they can remember."

Granny picked one of the flowers and sat down on the porch swing. I sat next to her. She pointed to the center of the flower.

"The corona is this circle of filaments in the center of the flower that represents the crown of thorns. Inside of the ring are stamens with five anthers which represent the five wounds that Jesus suffered. The three purple stigmas represent the three nails that the Roman soldiers used to nail Christ to the cross."

"How sweet," I mumbled and grew pensive.

"My little granddaughter, all grown up and a married lady now."

I felt the burden of sadness at the thought of my husband overseas fighting in a war that cloaked me with uncertainty and filled me with terrifying thoughts I could not control. Granny must have noticed the change in my eyes.

"What's the matter, dear?"

"Oh, everything, Granny. I lost our baby when I miscarried, and I can't stop thinking of who our child would have been."

"I know, child," Granny said, patting my hand. "We don't understand these things, but I do know that each of us belongs to God. We do not belong to ourselves. We are only the caretakers. This baby came through you, but he or she belongs to God, and He is in charge of life and death. We are not the god of ourselves—that thinking is in the devil's territory. I know loss is so hard to take. I lost my twelve-year-old son Herbert to pneumonia. His death broke my heart. He was born in 1900 and passed away in 1912."

"Oh, Granny, I didn't know that. I am so sorry. What was he

like?"

"Oh, my little Herb. He loved the sky when it was blue or gray. He loved the breeze that blew around this porch that made his paper airplanes fly. He sat right here and practiced folding paper to make different shapes of airplanes—experimenting with which one flew the best. In fact, he had a vision when he was about eight years old of an airplane mounted with guns and firing. This happened around 1908. What's amazing is that airplanes were not used by the military until 1911 or 1912 and only for reconnaissance or mail. It was later in 1915 that guns were mounted on the first plane and then became a staple of our Air Force."[7]

"Wow," I said, surprised. "Makes you wonder how he could time jump like that."

"I have no idea," Granny shook her head.

I slumped into my grandmother's arms.

"I'm so afraid of losing Larry, too. I'm so afraid he may not come home. It shakes me to the core. I love him so much; I don't know if ... I couldn't bear it. I prayed so much for him to come into my life, and I felt that God led us together, and now, I am so worried that God will take him away from me," I blurted out, sobbing.

"Aw, the life of the wife of a soldier is not easy," Granny said, patting my hand and putting her arm around me. "Sometimes it is easier to be in the big middle of it than to let your imagination go wild. I believe in prayer, and I believe in miracles. Keep praying for him, dear, as will I. We have a history of patriots who have fought for this country. Your great-grandmothers went through the same terror. You know the picture hanging over the old pump organ in the living room?"

Granny gestured toward the living room door. I nodded.

"That is John W. Hanchey and his wife, Eveline Arthur Hanchey. You were named after her. Eveline Arthur's father was the grandson of Captain Benjamin Arthur, who fought in the Revolutionary War. Let me tell you about Captain Benjamin Arthur. When he was a boy, his parents settled in rural Virginia. Almost all of Virginia was rural then, so he was quite far from civilization. His uncle settled not too far from their farm. One day, his mother asked young Ben to take a sack of corn to the mill to be ground into cornmeal. So, he mounted his horse and balanced the sack on his saddle, and took off. When he returned, Indians had raided their home, killing and scalping his Mama, Papa, and sisters. His mother was still alive but bleeding to death. She said to young Ben, looking into his stricken face, 'Son, I'm proud of you, and I love you more than my life. Protect your soul; don't let this turn into hatred. Forgive them.'

"Ben went to live with his Uncle Thomas and ended up marrying his cousin Ann many years later. When the Revolutionary War began, he signed up and eventually became a Captain and fought under General George Washington. We owe a lot to the patriots who were convinced of their right to govern themselves and not to be under British control. Captain Benjamin Arthur came home a wounded man and lived out the rest of his meager days with his wife and family. We have many other patriots who settled this land and fought for our nation. They were brave settlers and soldiers. So, war is not just the soldier's burden, although he takes the brunt of it; war is also about the courageous wives who agonized over their own thoughts, kept the home-front going, stood by, and prayed for their men. When the men did come home, they were there to heal them, care for them, and to love them back to life so that they could be whole again."

Granny paused. I thought about hate, and I asked myself if I hated war. Yes. Did I hate the Japanese people? No. They were fighting because their government made them. Their government was the aggressor. We were fighting to defend ourselves and our country.

"When will all this war and hatred stop?" I asked.

"You're asking about time by using the word 'when.' Time is the great limiter, the great tyrant that limits our thinking to the here and now. When your conscience is involved, you defeat time because it teaches the wisdom of the soul where it is in the realm of infinity, limitless. If you are asking about this war, World War II, I don't know. If you are talking about all wars, I doubt that they will ever stop. People keep thinking of creative ways to kill. It is part of our fallen nature. Hatred is brought to life by self-pride, greed, and selfishness. The notion that everything is all about oneself. Not until people recognize that God exists and that all sin defies the authority of God will hatred stop. Until then, we have to fight the serpent daily."

"With all this fighting, I sometimes think we are like fleas fighting over who owns the dog," I joked. We both chuckled. "Granny, how do you know all of this?"

"Oh, I would say that it comes from prayer and meditation. Music, too, because music can praise God. When I work in the garden, cook, or hang clothes out to dry, I think and pray. It's the solitude I enjoy. This world is bent on making things that speed through in the hopes of saving time. Fast cars, radio airwaves, and the telephone. They are all good, but it takes you away from the outdoors, the observation, and getting into the rhythm of the natural world, where meditation becomes easy. It brings me such joy and peace, especially when I get a glimmer of insight that can only be from God."

"Thanks for sharing that, Mama," I said.

"What is so sad is that Granny was talking about her grief, and in about ten years from that time, she would lose her husband, my grandpa. He was killed in a horrific car accident near Oberlin, Louisiana," Mama said.

"Mama, I remember that. I was about five years old, and we were driving to Dry Creek from Lake Charles. Daddy stopped at a red light and bought a newspaper. You read the headlines to us, and it was about the horrific crash that killed three people," I exclaimed. Mama nodded and said, "We were going to Granny's for the visitation, which was at the old homestead. The casket was in front of the fireplace. Closed, of course. We sat quietly together; I thought about the burdens my family before me had to carry. It put my small problems in perspective.

India 1943

"After we married," Daddy said, "I was sent to Camp Van Dorn, Mississippi. In September of 1943, the call came out after the Quebec Conference of August 1943, where President Franklin D. Roosevelt and General George Marshal requested volunteers of a 'high state of physical ruggedness and stamina' for a 'dangerous and hazardous' secret mission. Nine hundred and seventy men volunteered from jungle-trained soldiers based in the continental United States. I was one of them who had jungle training in the Panama Canal Zone from January 1942 to January 1943. They promised an early out—ninety days.

"I had only been married to your mother for six months, and she had recently miscarried, so I wanted to get back to my bride as soon as possible, so I volunteered. In no time, I was saying my goodbyes and boarding the SS Lurline out of San Francisco.

"I sent the following poem to my Mama before I shipped out. Years later, after Mama passed away, we dispersed her things; I found the poem in her Bible."

THE PRICELESS LETTER

I, a soldier, might write a hundred letters
To my sweetheart that I adore
And tell her in every missive
That each day I love her more.
I will tell her she is beautiful
In pages of praising lines
And mention she has the face of Madonna
With beautiful eyes that are dreamy and kind.

If I had the pen of a composer
And wrote at the end of each day
I would always end by saying
You are my brightest ray.
But the pages that are more welcome
That lights a spark in her gentle breast
Is the letter to my darling mother
From her soldier boy she loves the best.

I see her in the same old rocker
Where she held me long ago
And praying God to guide me
With a voice that's soft and low.
Regardless of my spelling
Or whatever I might say
The letter is a memorial possession
From her soldier boy far away.

Yet the letter of all letters
Is to my aging mother at home
For her anxious mind will always follow
Wherever I may roam.
So, all you fellow soldiers
Don't fail to write your mother dear
For she receives your most welcome letter
With trembling hands and a trickling tear.

-Sgt. Larry W. Stephenson

"The *Lurline* had been a luxury cruise steamer for the discerning and monied travelers that cruised to Hawaii and Australia with stops along the way. When WWII started, it was converted to a troopship. A friend of mine gave this old brochure about the *Lurline* to me; Evelyn kept it in her keepsake box. The *S.S. Lurline* was designed for 700 Passengers and 350 crewmen. It was 632 feet from stem to stern and 79 feet wide and could travel at a speed of 22 knots. The rooms were furnished with bunk beds made from pipes that were welded together. For the most part, they were stripped of luxuries that passengers expected during that time; however, the walls and wallpaper were intact. I think I remember the dining room walls were painted in a beautiful blue-green, the color of the sea when you are near shallow water. It also had wood appliques that were painted gold. I had never seen such luxuries like inlaid wood in the smoking room or the beautiful murals painted on the walls throughout the ship. The brochure's advertisement boasted of the *S.S. Lurline* in its heyday: 'The dining salon, a room whose urbane beauty delights the connoisseur as keenly as its cuisine enchants the epicurean' and 'spacious, sun swept, and sparkling with life...'[8]

From San Francisco, 960 volunteers from the Caribbean Defense Command, as well as 970 volunteers from the Army Ground Forces, boarded the *Lurline*. We steamed to the harbor in Noumea, New Caledonian, on October 1 to pick up over 674 battle-experienced veterans of Guadalcanal and the Solomon Islands.[9] Some of the Pacific veteran soldiers came out of stockades in exchange for their freedom. These guys became known as the 'dead-end kids,' which was a spin-off of Sidney Kingsley's Broadway play *Dead End* which was made into a movie. The original dead-end kids were a group of New York actors that were notoriously spunky and bad. So, the group of irreverent guys with a happy-go-lucky attitude that fought in Guadalcanal, New Guinea, and New Georgia volunteered for this hazardous and dangerous mission, as if Guadalcanal, New Guinea, and New Georgia were not already hazardous and dangerous," Dad said facetiously.

"These guys were creative and fierce fighters who had the attitude of nothing-to-lose because many of them grew up as street kids from larger cities. They hated the regimental life, the marching around, saluting, and stuff like that. They were kind of like the guys in the movie *Stripes* with Bill Murray."

I smiled, thinking of the craziness of that movie. "How do you know this stuff, Dad?"

"After many years of suppressing anything to do with the war, someone from Merrill's Marauders Association contacted me, and I started going to the reunions. There were several books out written by officers, and I read everything I could get my hands on. When you are a peon, you don't know the big picture. I wanted to know the big picture.

"Let's see, where was I? While onboard the *Lurline,* we had instructions on what to expect, scouting and patrolling, first aid, how to survive in the jungle, and all that prepared us for jungle combat. We learned about everything but where we were going. We had PT as well as hand-to-hand combat practice. In our spare time, we loafed around the ship's former bar or on deck, smoked, and talked to the guys that had volunteered from Guadalcanal. We were in awe of their war experiences and chewed the fat about guns, weapons, and fighting a shrewd and sneaky enemy. The main thing we had in common with these guys was that we were all angry over the bombing of Pearl Harbor and the deaths and casualties that incurred. Although we and everyone in America were pumped on patriotism, I knew the risks. I was ready to unleash my anger on the Japs.

"The food on board the *Lurline* was great and made us think that things wouldn't be so bad. Little did we know. They even had music over the ship's PA system; however, I noticed one song played repeatedly, and I became suspicious. It was the 'Song of India.' There were no words to it, and the only reason that I recognized it was because my mother played it on the piano.

"After our stop in New Caledonia, we went to Sydney, Australia, where we stopped for supplies. We stopped in Brisbane, where 55 volunteers from General MacArthur boarded. General MacArthur was supposed to furnish 274 volunteers, but that did not materialize.[10] We continued our journey through the Indian Ocean to Port Fremantle, at Perth, Australia. That's a lot of nautical miles.

"The days spent on the water were not agonizingly boring days, but they became a microcosm of my life: a day of my life that I had time to think, wonder, and reflect upon. There was so much water that I couldn't see any land. I watched the swells like they were the gentle heartbeat of the ocean, but when the weather turned, the ocean became an anger that was voracious with power. I began to look for changes: the light on the water at different times of the day, the movement of the clouds, the way the bow of the ship powered through the water. I looked from the stern of the ship and saw how the wake slowly dissipated back into the vastness of the water. There was a rhythm in the ocean. When we docked in Perth, the ebb and flow of the tide made it more noticeable. The rise and fall of the waves made me think of the ups and downs of my life in relationship to the sea. The rhythm and timing of the ocean and the stillness of its depths spoke to me of the truth that this was the way of all creatures on land and in the water. I had time for introspection and often wondered what the future would hold and if we would make a difference. I knew I was angry enough to kill the enemy. My religion taught me to 'turn the cheek' and not to kill. Surely that could not be translated to mean not to protect the innocent, not to protect the citizens, and not to protect our country. If that were the case, we would become nothing more than empty cans that are kicked around, worn down, and good for nothing. The bombing of Pearl Harbor became the 'shot heard round the world,' and the U.S. entered the war.

"Little did I know that those days of introspection would turn into days of exhaustion and days of extreme thirst. Yet, they were my days, even though the Army owned them since I enlisted. We finally steamed through the Indian Ocean and up the Arabian Sea, and at last, we docked in Bombay, India. It was a long and tiresome trip. When we disembarked, I was so thankful to be on solid ground that I could have kissed it. We arrived on October 29, but our battalion didn't disembark until October 31, 1943. Now as I look back, Halloween was a precursor of the horrors we were getting into.

"Once we docked in Bombay, we marched single file through the dusty city filled with bullock carts, tangas, autos, and throngs of people. The sights, scenes, even the smell of curry were new to me. It seemed so far away from war. People were going on so casually with their lives and the business of buying, selling, and bartering that I wanted to scream, 'Don't you realize that war is on your doorstep? Your neighbors in Burma are fleeing from war, and we are going in to stop Japanese imperialism! We are going to war! If we don't stop them now, they could take India, go through the Suez Canal with their ships, hook up with the Germans in Europe, and eat the world's lunch.' The Germans and the Japanese were allies. In fact, the Germans considered the Japanese 'honorary Aryans.'[11] The major objective was to oust the Japanese, but our immediate objective was to take northern Burma from the Japanese so that there would be a straight shot for supplies from China over the Ledo Road, and the pilots didn't have to fly the hump over the Himalayas anymore. My mind was exhausted from the tirade. I kept on marching, and a little Indian boy about three years old started marching beside us with exaggerated steps and looked at me with big brown eyes and a happy grin. Funny how a smile and the innocent antics of kids can turn your mood in a heartbeat.

"We marched to a railway station operated by the Great Indian Peninsula Railway, where we boarded a train that lumbered down the tracks like a hulking prehistoric creature, snaking around mountains, making hundreds of stops. The train was slow as watching the clock when you can't sleep. The first thing I noticed was the poverty. When we came close to a town, desperately poor natives that had camped on the tracks, carrying their belongings tied in a piece of cloth, came out begging and stretching their skeletal arms to the windows. The level of destitution shocked me: the swollen bellies, sunken cheeks, the dull eyes. These were human beings acting like a swarm of flies; that was desperation. These human beings had hearts, minds, and souls. When I broke from my shocked stare, I handed them crackers and stuff from my rations. All the guys did.

"The natives tried to sell us anything and everything. We could buy things like tangerines, whiskey, chickens, and even monkeys. After the train was on its way, Lieutenant Weston, after observing the poverty, said to those within hearing distance something like, we are fighting for our American way of life, and if we are not willing to fight here and now, we may not have the comfort and peace of home to go back to.[12]

"The train carried us to the British post at Deolali, which was typically British with its impeccable parade grounds and brick buildings for officers. After one day, some of the guys approached the turbaned tea wallahs who hawked their tea by saying 'good British tea.'"

'Hey wallah, wallah, here's a rupee to say new words,' the GI said. 'Say, 'good British piss poor tea.' The GI coached the wallah, so he had the right pronunciation. Then, they went to another tea wallah hawking hot tea and gave him a rupee to say, 'Hey GI Joe, Good Betty Grable Tea.'

The wallahs hawked their tea with other irreverent words, much to the fury of the British. Of course, the tea wallahs had no idea what they were saying. The British were offended by the lack of respect for their traditions and lack of respect for British royalty, so the Americans were moved into tents off-post.[13] The funniest thing I saw was the dead-end kids from the 3rd Battalion trying to brew coffee using a brassiere as a filter. It gave a whole new meaning to A cup of coffee. After a few weeks of orientation and training, much to the delight of the British, we were loaded onto a train for Deogarh, where we trained for jungle warfare."

"That's hysterical!" I laughed.

"Yeah, we had to keep our humor to mask how scared we were. It also helped with the morale, the 'esprit de corp,'" Dad said.

"Continue," I encouraged.

"When we were at Camp Deogarh, we marched to different areas and practiced war games. We practiced infiltrating columns, attacking an airfield, attacks on columns, firing on range, scouting and patrolling trails, and swimming for non-swimmers. I could already swim, so I didn't have to do that."

"Dad, I am surprised that anyone would be accepted in this outfit if they couldn't swim," I said.

"I thought so, too, but maybe they needed the volunteers, or those guys slipped through the cracks," he mused. I remember before Christmas in 1943, the dead-end kids went AWOL and commandeered a train to Bombay and various Indian towns, blowing their money on every conceivable pleasure, from drinking to women. They came back to camp broke and three sheets to the wind. They were all busted, but they didn't give a hoot. On New Year's Eve, they fired their weapons into the sky, trying to shoot the stars, and howled at the moon. I sat outside my tent looking at the moon and remembered how romantic the full moon was in Natchitoches, Louisiana, where Evelyn went to college. That's where I fell in love. Looking at the waxing moon in India that night, I felt a dark and eerie anxiety that was juxtaposed with the brightness of the moon and the jubilant and drunk soldiers. I didn't know then how hard it would be to see the moon again because of the jungle canopy. It felt like the light had gone out for me, and I was entering the darkness of the unknown. Thinking of Evelyn was a way for me to squash my fear of the unknown. I had to discipline myself and not let my imagination run away with me. I pulled out paper and pen and began a letter to Evelyn."

Friday, December 31, 1943

Darling,

I'm sitting here looking at the moon and wondering if you were doing the same. I miss you so much that it makes my heart squeeze. I hope you are well. Take care of yourself. You know I can't say where I am or anything about this secret mission, so don't trouble yourself by worrying. It is New Year's Eve, soon to be 1944. I hope this is the year the war ends, and I can come home to you. Last year about this time, we were planning our wedding.

Remember Darling, how your dad borrowed enough gas rations, so we could go to Jennings, Louisiana on our honeymoon?

That was as far as the rations permitted us to go. I will make it up to you someday. I loved it when you hummed your Chopin music and played the piano on my back. You are so funny. Keep playing your music my love because you are incredibly good. Besides, it may take your mind off the fact that we cannot be together.

Happy New Year's darling. Save it all for me. An ocean of love and a kiss for every wave. Pray for me. Take care of yourself and give my regards to your folks.

Love,
Larry

March to Staging Area 1944

"In the first few days of January 1944, we learned that our new moniker, other than Shipment 1688 A, B, and C was the 5307th Composite Unit (Provisional). Not exactly a fighting title or morale booster, but more like an asterisk of omitted information tucked somewhere in the footnotes or the index that maybe no one would notice. Later, I found out that our code name was Unit Galahad, but James Shepley, who was a *TIME Magazine* correspondent, coined the name 'Merrill's Marauders.'[14] There were always rumors and more rumors, so they had to be taken cautiously. We had heard that British General Orde Wingate was to be the commander; then, we heard General Joseph Stillwell was to take command. It was like some kind of pick-and-choose shell game. All these wishy-washy, last-minute decisions certainly eroded our confidence in the organization of this outfit.

"I was assigned as a Sergeant in the 3rd Battalion, Orange Combat Team commanded by Lieutenant Colonel Charles E. Beach. He was a veteran of the Solomon Island Campaigns, which, luckily for us, generated a lot of confidence among the men. Years later, after the war, I heard that Colonel Beach committed suicide. Major Lawrence Lew was the commander of Orange Combat Team, which included Company L, which I was in. After learning who we were on paper, we were off to the battlegrounds to show the Japs who we were as fighters.

"January was supposed to be the coldest month of the year, but it was warm, dry, and dusty during the day and cold at night. We were heading out to our staging area. When the train pulled out from the station, about ten Indian men and boys hopped on the last car and held on for the ride. Some of those rascally boys would run through the cars grabbing bags and pitching them through the windows where their cohorts ran along the side of the train and caught them. The train was crude, especially in 3rd class: box cars, wooden seats, sleeping on the floor, hole in the floor for a toilet, windows open during the day, closed at night, K rations, and cockroaches careening to compete for food droppings. The train was woefully slow—fifteen to seventeen miles per hour. A week of being bumped around lighted a fire under our anger temperature gauge. Here we were, the rugged warriors, with headaches from smelling coal smoke from the smokestack, sweat, puke, and the hit-or-miss toilet hole called the 'convenience compartment' located in the fore of the box car. For God's sake, we were supposed to be fighting men, but it sure did bring out the wuss in us. One of the dead-end kids bought a cow and brought it into the box car. Anthony was a butcher before the war, and he butchered it and cooked it right in the box car. We feasted on beef steaks which were a lot better than K rations. I hope the Indians weren't too upset that their holy cow was diminished to a bunch of bones and carcasses thrown out of the box car door while we were moving and strewn along the railroad tracks.

"After we got off the train, we rode a paddle-driven ferry up the Brahmaputra River for three days, which was a more pleasant way to travel, except for sleeping on the deck of the stern at night and freezing to death. The next night, I slept on the starboard side, mid-ship, which was much warmer. We had covered about 1200 miles in ten days and had another 250 miles to go. We were on the last leg of the journey when we boarded the Northeast Indian Railways that ran from Gauhati to Margherita, the hill station near Ledo, in the state of Assam, India.

"The only entertainment was baboons swinging through the

trees until we came upon naked gals bathing in the river. The boys started yelling obscenities and whooping."

"Sorry to be ungentlemanly," Daddy said, "but one guy yelled out the window, 'Hey Hotsie-Totsie, you want a ride on my doodle-dasher?'" He paused to let it sink in.

"Doodledasher?" Mama said lamely and looked at me.

She and I broke the silent pause with a burst of hilarious laughter. "All the guys laughed too," Dad said. "Someone said, 'We ain't never heard of a doodle-dasher.' 'That's what my Pops called it,' he said in his defense. Well, you guessed it. His name became Doodle-dasher, shortened to Dash. I never knew his real name."

"You know, the expendables were not supposed to write notes or keep diaries about the campaign, especially because it was a secret mission," Dad said as he opened his diary slowly, "That was relegated to the officers at headquarters."

"I read in one of your Merrill's Marauders books that Merrill himself ordered that no notes or records were to be kept about the mission.[15] That's why your diary is so important," I said.

"True, but I didn't know the big picture like the officers. My diary is from my perspective as a non-com. I'm going to skip much of January and ten days of February," he said, "because we were training, and many of the entries are simply 'usual camp duties.' In February, we traveled by train. I will begin with the march to the staging area."

> *February 11 [1944]. Monday. En route to new station. Left bivouac area with 4th plt [platoon]. My squad is point for 5th dispersal group. Moved out at*

8[:]20 p.m. Tired and sore. Rained for 19 miles. Sleep at 4[:]30 [a.m.] Up at 9[:]30 a.m.

February 12, [1944]. Up 9:30 [a.m.] In bivouac area on Ledo road. 7 miles fr town of Ledo. Washed up. Rested. cleaned M1. Rained. Started to move out again at 6:45 [p.m.]. marched to 2nd bivouac in shacks. Arrived 1:30 a.m. Very tired and sore. Rained. Uphill and downhill. Rough 16 miles.

February 13, [1944]. En route to new station. Arrived 1:30 a.m. Miles marched 16. Left Bivouac area 5:30 [a.m.] Feel better. Arrived 3rd bivouac area 11:45 [a.m.]. Feel a little better. Slept in barracks next to Lt. Bert. Air mattress is swell. 2 meals per day rough. Miles marched 10. Left camp at 5:30 [p.m.]. 2nd dispersal group under Sgt. Mulligan. Arrived at 4th bivouac area at 8:30 [p.m.]. Miles marched 2.

February 14, [1944]. En route to new station. Saw Melvin Douglas in show. Had grapefruit juice last night. Slept cold. Still on Ledo Road, lots of traffic. Cut my squad's hair today. Very poor chow. Rested. Tonight, we leave camp at 5:30 [p.m.] and will be on Burma Road at 11:30 [p.m.]. March wasn't too rough. Covered 12 miles. My feet are not in 2 bad shape. Arrived in Tanchu rest camp. Had movie and good coffee. Slept well. Rained like hell. Men got wet over the Big pass. Cold as hell.

February 15, [1944]. En route to new sta. It was rough last night & cold. Left bivouac on mt at 5:30 [p.m.]. My squad is on rear of 2nd dispersal group. Miles *marched 12. Saw movie. Coffee and tea. Bed at Tanchu. 59.25 miles from Ledo. Washed up during day and wet shoes and towel. Read book and then left out again at 6:30 [p.m.] in 4th dispersal group. Enroute to 7th bivouac. Am wearing jungle boots tonight. Try them out. They didn't work so well. Arrived at bivouac area at 11:30 [p.m.]. Tired as hell. Distance marched 12 miles. Bed 1 p.m.[?]*

February 16, [1944]. Up at 10 a.m. Had poor breakfast. Cold last night. Slept 3 hours then cleaned up. Have bad knee. Left cp at 6:00 p.m. New cooks and mess sgt Moved out at 8:20 [p.m.]. Arrived new bivouac area at 9:45 [p.m.]. Bed 10 [p.m.]. Crossed river.

February 17, [1944]. Up early read book. Action at Aquilla. Washed up. Good Breakfast. Will move out early. Moved out 3 p.m. Fourth dispersal group. The hill was really rough. Had tea at negro camp.

February 18, [1944]. Up 8 a.m. on top of hills. Clouds below us in valleys. 2 more miles uphill. Had fair meals. Everything tactical from now on. Moving out at 3:15 [p.m.] for 7 miles. Left out and marched 8 miles instead. Boy, am tired. Arrived at 9:30 [p.m.]. Bed and sleep. Up hills pretty rough. 6 miles downhill.

"The beauty of the country was stunning. It felt like we were on top of the world. We walked through the clouds to a height where we looked down on them. It certainly could make you reflective and nostalgic, but sooner or later, your legs began to cramp, and it was hard to take a deep breath, and you were back into reality. It was hard to see the mules and horses struggle. That's where I first met Red. That was my name for her. I always named my pets names that were opposite of their physical looks because it would make it harder for them to be stolen. It would be common for a thief to say, 'Let's go, Blackie,' or 'Let's go, Beauty.' She reminded me of the horse in *Black Beauty,* except she was straggly. She balked when the muleskinner tried to get her to climb higher, so I and some of the other guys made sure her load was perfectly balanced. We cut out steps up the mountain with our entrenching tools to give the animals better footing. Finally, I stroked Red and talked low into her ear, and gave her a sugar cube. Sure enough, J.D., the muleskinner, and I were able to coax her to climb. I made sure there were no pythons that looked like logs that scared the guys in Lieutenant Weston's I & R platoon."

"What happened to those guys?" I asked.

"They had an ornery mule that wouldn't climb the hill. So, the guys braced themselves against a log to push the mule from its backside. No sooner had they started when the log began to move. Not a rolling move but a side-winding move! It was a giant python! The muleskinners jumped about six feet straight up into the air. The guys around them got a good laugh, but it scared that mule. She could smell it. The python slithered away into the underbrush. They could have shot it, but they didn't because it wasn't a threat. Besides, all the creatures, including us, were trying to survive this war."

> *February 19, [1944]. En route to new area. Up early. Rested and ate breakfast. Moved out 445 [a.m.] 4th dispersal group. Arrived 10:30. Mahmood and rest had 7 a.m. guard. Ate poor supper at bridge. Bivouacked on top of hill. Gorretti slept in Negroes tent and drank beer.*

February 20, [1944]. 8 miles. Up early. Down to breakfast at curve in road. Goddard fixed me up to sleep in native shack. Swell. Breakfast and tea in. Rested well. Mahmood. Goretti. Roger & Bert say should be in tonight. Just about to Foothills. Hurray. Moved out again at 11 a.m. We are now the fifth dispersal. Went 11 miles and boy was it rough. Mud Rain and everything. Bivouacked beside road between big logs. Rested & it rained like hell. Arrived at 3 p.m. Got coffee and extra chow at kitchen across bridge by Chinese camp. Rained.

February 21, [1944]. Up 7 a.m. Everything is wet and we are cold. Air mattress was swell but is heavy. Lt. gave me musette bag. Have a very heavy pack. Left bivouac area 10 a.m. Marched all day. Boy, it's tiresome. I'm sore. Shot one mule. It's wore out. Just 5 miles from Japs. Marched 18 miles. Arrived 9 p.m. at paddy field. Dead Japs all around us. Chinese too. Colder than hell and tired.

"You shot a mule?" I asked.

"Sometimes, it was the only humane thing to do. They would get sores as big as my fist from the poorly fitted pack saddles and the heavy load. Many were not healthy, to begin with," Dad explained. He continued, "On February 21, I remember marching and coming into a paddy field and seeing the first dead Chinese and Japanese soldiers lying in various stages of action stopped. They were in distorted positions with grotesque expressions—mouths open, exposing yellow teeth, sometimes shiny gold, against swollen black tongues. Many had maggots in their mouths. Their skin was mottled, grey, or black. A few were a bluish color; perhaps they had been alive longer. Their blood and entrails were spilled out around their bodies. Most were on their backs, obviously from the power of a frontal hit that propelled them backward. The stench was sickening."

> *February 22, [1944]. Up 7 a.m. Rolled packs. Bivouacked 600 yards in jung, cleared out place for our squad. Bath and washed clothes. 11 [a.m.] BN is on 2nd watch near river. Dead Chinese and Japs all over the place. Given 10-1 rations. Found wild lemons. Have swell place for camp. Smuck is good man. Gurkha knife comes in handy. Bed Early.*

"Dad," I said, "can you elaborate a little more on that part of your diary?" "Well, let's see," he began, "once we got to the hill station, the tracks ended, and from there on out, we marched everywhere we went. The Ledo Road was a strategic, overland supply road to China, and it was part of the Marauders' mission to keep it open. The main port in Burma was in Rangoon. It was held by the Japanese, which left The Ledo Road, that swath of a single-lane road that American engineers were carving out of the jungle, as the supply route for us and our Chinese allies. There were barracks as such, some previously built by British and Chinese troops, but most were built to house the American engineers. We were grateful for a place to sleep and shower.

"In the beginning, we marched at night to keep a low profile. However, our cover was blown when Tokyo Rose, who was an English-speaking radio propagandist, was informed that American soldiers were spotted on the Ledo Road. She had a radio program that she used to demoralize and debase our soldiers with her wicked snipes, but most of the soldiers thought it was a joke. It was recorded in Japan and beamed out over the Pacific, and of course, by then, the Japanese in Burma were aware of the American soldiers. Now that the element of surprise was lost, we started marching in the daytime.

"It was crowded coming out of the hill station near Ledo, but when we crossed the border into Burma, the traffic slowed down except for the natives on foot. The rain made it a holy mess. We tried to cut steps in some areas, so the mules would have a solid footing, but some of them slid off the edge of the mountain to their deaths. That made me sad, and the muleskinners even sadder. We marched about ten miles the first day, sixteen miles the next, and when the mountains got steeper, we could only go about six to eight miles a day. For about twelve days, we marched to get to the area where our first mission was to begin.

"We finally made it to a rest camp where we had to take typhus and other shots. We laid around and got some rest. I read my book *Action in Aquilla* and then watched a Melvyn Douglas show. I was made Sergeant in the 4th platoon under Sergeant Mulligan and Lieutenant Burk. I took my duties seriously and was determined to get every one of the men in my squad back home. I remember when some of the soldiers in my squad and I sat around camp in a little patch of sunlight, cleaning our automatic weapons."

"Three Rs: Rust Ruins Rifles. If you don't clean and oil your rifles every day, you're screwed," I said. "They will jam, stick, and not shoot straight."

"Yep, my best friend from here on out," Smuck said. "Your best friend is your short arm," Mac said, laughing.

"Well then, protect it as good as you do your short arm, and you will be okay," I said.

I paused, and everyone silently cleaned their guns. I wondered about killing a human being and whether we would survive.

"You know, guys," I said, "we are going up against the 18th Division of the Imperial Japanese Army. Those were the Japs that raped women, children, nuns, and anything that walked in Nanjing, China, a few years ago. They massacred 300,000 people. They even threw babies in the air and caught them on their bayonets. Now you know why these natives are getting the heck out of here. Another thing you need to know about the Japanese warrior is this: The Bushido code is the 'way of the warrior' in the Samurai tradition. They are taught that they are already dead, so they have nothing to lose when they go into battle. The Shogun is the Commander in Chief and fights against the barbarians. They believe he is descended from one of their gods. So that's one of the reasons they will fight to the death."[16]

"How do you know this stuff?" Taylor said.

"I read a lot of books, and I read the newspapers. My Mama was a teacher in Texas until we moved to Arizona. She made us read."

"Those yellow-livered cowards haven't gone up against us yet. We are one mean, fightin' machine," Gorretti said.

"And angry as a double dose of burning hell," Smuck said.

"They must be crazy if they think raping and killing women and children and old men isn't barbaric," Mac said. "If they had raped my girlfriend, sister, or mother, I would fight like a cool, cold killer with no mercy."

"I'm gonna shoot their faces and their peckers off. They deserved to get it blown off," Taylor said.

"That's why we have to stop these no good, yellow-bellied devils," I said. "You gotta be careful that you don't let your anger overshadow common sense and the training you've had. Don't let it make you do something stupid. Besides, our weapons are superior to theirs."

"We moved out the next morning and marched nineteen muddy miles in the rain. The next day sixteen miles. We passed a lot of native women, ragged children, and old men carrying everything they owned on their backs or in baskets attached to their heads with a head strap that went across their foreheads. It was a sight to see. They were fleeing the Japanese, and we were going toward the danger. They stepped off the trail and rested so we could pass. We continued climbing the rugged Patkai Mountain Range, and soon we were above the clouds about four thousand feet up. Forests of huge teak, pine, mahogany trees, and bamboo towered above us like a forest of giants. It was beautiful up there, a taste of heaven. I remember the refracted sunlight shining down from heaven casting everything in a crystalline light. It made you feel so close to the creator. I would have enjoyed it more in different circumstances, but I had to snatch the beauty and memorize it because it was cold, and the reality of one mudslide to death preyed on my survival mode. My wet pack felt like I was carrying a load of bricks. Climbing up the hill was okay; it was coming down that was hard because of our heavy backpacks. I felt sorry for the mules. They were worn out, too, and had sores from the heavy gear on the ill-fitting packsaddles. Some of us had to help the muleskinners push the mules from behind to get them up the slick slopes. They were shot when they wore out and collapsed. After the War was over, I found out that the ship that carried 500 big, sturdy mules that had been picked in the U.S. and sent to India for our use was torpedoed by a German U-boat (U-178). They were on the *Jose Navarro* when it was lost in the Indian Ocean on December 27, 1943.[17] Our muleskinners were to work with them before we started on our mission, but those poor mules became a feast for the sharks. Horses from Calcutta and Australia were rounded up and sent to us, but they were not really suited for jungle transport. The mules would eat bamboo leaves, but the horses would not. They dropped grain for the horses, but when we didn't get an airdrop, they had nothing to eat.

"Everywhere we marched was hostile, from the slippery slopes to insects, mosquitos, microscopic enemies, animals, snakes, and Japs. Even the constant rain was hostile. The only living creatures that wanted us there were the Kachins. They hated the Japanese as much as we did because the Japs threatened their homes and villages, raping their women and burning their men alive. Our Kachin guides were invaluable to the success of our mission. Their bodies were well-adapted for climbing rugged mountains. They were long-waisted, short-legged men with incredible stamina developed from going up and down those mountains all their lives. Their short, sturdy legs gave them balance, and it made me think of the environment's impact on their bodies. The Jinghpaw, one of the subsets of the Kachins, were headhunters at one time, but the Christian missionaries gradually changed this practice, and they became a peace-loving, simple people, except that they wanted the Japanese extinguished as much as we did. Their primitive weapons of crossbows and blowpipes were their only defense; however, their guerilla warfare and hit-and-run tactics were superb. Their knowledge of roads, trails, and even winding elephant trails that we could use to disappear into the jungle was invaluable.

"Since we got off the train outside Ledo, where we stayed at a rest camp, we had marched for seven days on winding roads and trails. I thought of Evelyn as I lifted one foot after the other. I wrote letters and poetry to her in my mind, and at some point, I wrote them down."

LET ME KNOW

"I thought that you would like to know
that someone's thoughts go where you go.
That someone can never forget
the hours spent since we first met.
That life is richer, sweeter for
such a sweetheart as you are.
And now my constant prayer will be
that God may keep you safe for me,
and that wherever in peace or war I go
that you will love me and let me know."

-Larry W. Stephenson

Daddy looked at Mama with a sweet smile. Mama gave him an equally sweet smile. Something passed through them in those smiles. Those were the moments and memories where love was almost tangible.

"Before you knew it," Dad said, "we had marched nearly 140 miles over the Pangsau Pass border crossing from India and over the Naga Hills, which was a part of the Arakan Mountains that separated India and Burma. It was quite an effort to reach our staging area. Everything was tactical from there on out. On February 22, our I and R Platoon created the head of the serpentine column that guided the Third Battalion and scouted a bivouac area about 600 yards into a jungle clearing. I think it was our forward staging area which I was not aware of back then. We bathed in the river and recuperated from the march. Some natives treated us to cooked chicken. What a treat. You will never know how much you value food until you don't have any."

Mama served us coffee and a piece of cake, and we paused as we ate. Dad's memory of cooked chicken reminded me of an incident when I was a child. I wasn't in first grade yet, so I must have been about five years old. I remember my sisters and I getting dyed chicken biddies for Easter. They were sweet chicks dyed pink, blue, and green; however, when they grew up, you could no longer see the dyed colors. We got a rooster and kept them in the backyard. They multiplied and multiplied. Daddy was never keen on having chickens, but he went along with it until they started to get out of the chicken wire fencing regularly.

Mama called Daddy at work around noon and told him the chickens were out again. He came home in a fury. He didn't say a word—he just ran after the chickens in his white shirt with his tie flapping. He was incredibly fast and would grab one and wring its neck, then grab the next, then the next. The chickens ran around without their heads until they collapsed.

Daddy's white shirt looked like it had the measles from drops of splattered chicken blood. Then he put his coat on over it, straightened his tie, and went back to work. Mama and I stood by my Radio Flyer watching. I know my mouth must have been agape, but Mama was cool, calm, and collected. She piled the chicken bodies on my wagon, and we sat outside and plucked the feathers. We didn't have a freezer, so Mama went to a meat locker business that was next door to the A&P grocery store on Ryan Street. She rented a locker and froze the chickens, and we ate fried chicken, stewed chicken, barbequed chicken, baked chicken, and chicken and dumplings until we were sick of chicken. That evening, Daddy came home with a small jar of peacock blue ink. He made a quill out of one of the chicken feathers and showed me how to dip it in ink and draw and write with it. As I reflected on the incident as an adult, I realized that uncontrolled rage was a symptom of Post-Traumatic Stress Disorder (PTSD), a term that had not been coined until the 1980s. One of the most common terms to describe PTSD back then was shell shock, but battle fatigue was used as well.

I started the recording again, and Dad began.

"The noise of the jungle was at times clamorous with monkey and bird chatter, trumpeting elephants, and all the noises of the animals that call the jungle home. Sometimes the monkeys made a creepy, mocking kind of laughter as if they knew something we did not know. Then other times, they were like look-outs in the tops of trees. They were very quiet, but if you watched them, you noticed several of them turned their heads while watching either an enemy soldier or other animals. You had to be aware of the animals and birds around you. I never realized how deeply I could listen to the universe when the jungle was silent, when even a leaf was not stirring.

"If you are silent here at home, you hear white noise from the refrigerator, clocks, air-conditioning, electronics, and even the settling of the house. It becomes a conglomeration of vibrations and frequencies that emit a hum. Listening in the jungle became an anticipatory action, determining if the noise was one of danger. I had to stop myself to determine other noises that I could not hear, like the heartbeats of our soldiers, the heartbeats of animals, the flap of birds' wings, the movement of all living creatures, and even vegetation that emits frequencies of noise. When I had my helmet on, it was hard to determine the direction of the noise. That's why I preferred the soft jungle hats; besides, they didn't make noise when a limb hit your helmet, which would give your position away. I worked at listening to all of the sounds and vibrations that became the heartbeat of the jungle. I felt a commonality with these animals: we were all listening for danger, and like them, I turned this defensive listening into offensive listening. Opportunistic listening so that I could surprise the enemy. It was the first step to becoming fearless.

"I remember when I was leading my squad. We came upon a thicket where I heard noises—low talking, not in the rhythm of English—a language with a quick clip and intonation yet spoken softly. Naturally, I assumed that it was Japanese. When your life is at stake, you don't assume that someone is on a hike in the jungle. You assume it is the enemy. The jungle isn't a courtroom. "With my Gurkha knife in hand, I crept stealthily, being careful of every step to avoid the crack of a twig or brush in my path, and surprised a Japanese soldier on guard duty. His back was turned to me, and he was looking up into the trees when I sprang upon him from behind. I put my hand over his mouth and lifted him off the ground. His arms and legs pawed the air."

Dad made a pawing motion with his arms.

"I was quick and fierce when I slit his throat. His blood felt warm on my hands as I waited for it to run out of him. When he was limp, I laid him down gently and motioned for the guys to come in. We crept up to the opening of a clump of bamboo where a small squad of Japs squatted on their ankles around a fire, cooking greens of some sort. In fact, their pallor looked green when they turned and looked at us. We took care of them quickly. But I'm getting ahead of myself," he said and looked at me, "Sorry."

I sat shocked in the wake of the raucous casualness of that testimony. I knew my dad was no dummy, and he was checking me out to determine if I was going to be able to go through with this. I didn't blink an eye, but I was too shocked to speak. Dad continued.

"I've often wondered, since the war was over, what that Japanese soldier's last thought was. I hope it was worthy of his death. What would my last thought have been if the timing had targeted me?

"Although the approach march to our staging area behind enemy lines was long and arduous, I think back and was glad for it. At the time, I thought, even though the top brass didn't think it was a feasible suggestion—probably because it was cost-prohibitive and would destroy the element of surprise, but why couldn't we have been flown in a few at a time, so the march and physical exhaustion would not have sapped our strength before the battles? They had their reasons. I suppose learning to listen in a hostile environment was lifesaving and acquired simply by doing it repeatedly and learning to focus, so in that respect, marching was a good thing.

"Deep in the jungle, sight was limited so much that you could not see around a bend in the trail because of the giant bamboo that grew one hundred and twenty feet high—one stalk could grow to twelve inches thick—or the six-foot, razor-sharp kunai grass, and trees that were sixty to eighty feet high providing a canopy above that nearly shut out the light. That was when it was important to develop listening skills. It was also a way to overcome being unnerved because we fear being blinded or being in the dark. Knowing that we are so dependent upon our sight, even when it is impossible to see, we can envision a path to safety, and we can calm our sense of survival. It makes us feel a little safer. This, as well as training that prepares for surprises, makes us calm and go about our jobs."

We opened our one box a day K rations and traded items like currency. "You can have my chocolate bar for your cigarettes," Al Mahmood said to all of us.

"Who wants the battery acid, aka lemonade, for dextrose?" I asked. We bartered and traded until everyone was satisfied. We weren't always rowdy and fun-loving. We had our serious moments.

"What do you think about intuition?" Taylor asked, looking at me. "Can you trust it?"

We all thought for a moment. "I'm not sure, but I wonder if your brain takes a picture of, say, the trail. Maybe that scene is like a still photograph in your mind. Your conscious mind is focused straight up the trail, and you don't see a rifle muzzle sticking out of the leaves, but your brain does, and you get an eerie feeling like intuition."

"Maybe," Mac said. "Do you think your intuition is God guiding you?" We sat with blank stares.

"Maybe we should ask 'the fighting preacher' Lieutenant Weston. He carries a small Bible in his pocket all the time. He would probably know," I said.

The next time I had an opportunity to ask Lieutenant Weston, a nice fellow who didn't mind a sergeant asking such a question, he came by our squad and spoke to the men.

"You have to communicate with God to know that He is real. He is knowable through constant communication. I believe that intuition is that still, small voice of God for those who seek Him. It takes an interior type of listening, and it gets easier with practice," Lieutenant Weston said.

That was weird, I thought because I had just been talking about listening to the guys. Interior listening expanded my thoughts.

Walawbum

"Other than a couple of small skirmishes, Walawbum was our first mission. It took us three days to march to the area where the operation was to begin. The main objective was to cut off the Kamaing Road, which was crucial for gaining control of the Hukawng Valley. We were to go around the Japanese front lines so that we could establish a roadblock on the Kamaing Road behind them. So much of what we did was flanking movements to get behind the Japs, create havoc, and blow up their supply dumps."

February 23, [1944]. At Bivouac area near Nainka, Burma. At edge of River. Slept, cleaned up. Put shelter up. We are moving out tomorrow. Turned in shortage of squad. Had 4 cups of good old coffee. Expect Bombing raid.

February 24, [1944]. On our mission. Up at 6:30 [a.m.] Cold as hell. Started to move out at 9 a.m. Held up until 11 [a.m.] Passed river where 900 Chinese were killed. Thru 12 Jap positions & marched 11 miles thru jungle. Was very thick. Arrived in bivouac area beside trail at 530 [p.m.] Coffee and meat and beans. Not bad. My squad had right flank.

"On February 24, as we moved through Jap positions, I stopped briefly to take a whiz. I thought I was alert, but I found out that I wasn't. I happened to look up while relieving myself, and there, suspended in a tree, was a human rib cage with a heart hanging out of its brokenness. It was covered with flies. The rest of the body must have been blown to kingdom come. We marched several more feet and passed a river where hundreds of Chinese and

Japanese had been killed. I didn't gawk this time at the scene of so many dead littered about, but I kept my wits about me—looking into trees and brush for the enemy. One thing that I noticed was that some of the Japanese had writing on the flys of their pants. I later found out that it was their name and the company they were in. I was so intent on staying alert that I tolerated the smells. Later, after I had time to reflect upon it, I realized how unprotected I felt after seeing so many of our allies, the Chinese, killed."

> *February 25, 1944. Up early 630 [a.m.] Jap planes flew all night. Had stew and coffee for breakfast. Moved out at 9 a.m. 4th dispersal group. 1st Bn C Co. All the old boys. I & R contacted Japs. One of our men shot thru nose. 11:30 a.m. 1 Jap killed. 8 miles bivouacked. Heard lot of mortar & BAR fire.*

"I heard that there were four dead Japs in the jungle," Smuck whispered to me, then paused while he put his face in his helmet and took a deep drag from a cigarette. Smuck was hiding the orange tip because we were not supposed to have any fires that may give our position away. Nicotine addiction was a real thing that the guys had to contend with. Most of us smoked because it made us feel less hungry.

"It was Werner Katz, the scout from Lt. Weston's I and R platoon, that got shot in the nose. Didn't do too much damage. Scary thing was that a Jap stepped out of the brush onto the trail with a friendly grin and waved to Katz. The men thought they were Chinese until they opened up on them. Katz and Weston unloaded on them—killed them, and knocked out the machine gun. Funny thing, Lt. Weston apologized to his men for having to kill those

Japs."

"Wow, they were lucky. What cagey enemies the Japanese were. Staying alert means life or death," I said.

We fell quiet, stretched out on the jungle floor, smelling the acrid humus after the last of the cigarette smoke floated away. Being quiet and having no fire at night was isolation. It was hard to get used to, but it was necessary. I felt bound to the ground under the jungle canopy. I have never felt more trapped, and I knew I would have to fight my way out. I felt incredibly sad and wanted to wail. I must have been a whack job to volunteer for this BS, I berated myself. I never voiced that sentiment because it would diminish morale. In my mind, I put that thought in a coffin and nailed it shut. I was still gung-ho; however, I was also lonely for home. I wanted to see the stars and know that there was a world out there—a world of propriety and courtesy, a world with my beautiful bride. I listened to the distant mortar fire, more determined than ever to get back home. I fell asleep on the cold, hard floor of the jungle trail engulfed in darkness, and the next morning was born into the light of day. I had to believe in the light to come because the evil that surrounded me threatened to separate me from God and those I loved. I was grateful to have another day to live.

February 26, [1944]. Up at 630 [a.m.] Moved out at 730 [a.m.] Nothing doing last night. Just mortar fire. Held up for 40 minutes by 20 Japs on patrol. Arrived at new area. Marched 11 miles. Set up defense on left of BN. 1 man KIA. Fixed chow. Short of it, guards say Jap patrols active.

"We marched eleven miles through the jungle, being careful not to disturb the branches and leaves along the way. Before we headed out, I always checked my compass because when you hold your rifle in a ready position, you don't have time to fiddle with anything else. I always oriented myself as to the direction we moved in. For the most part, I kept track in my head of the steps marched and tried to remember the left and right turns we made because we rarely saw the sun. When the sun was out, I thought of the cardinal directions. I knew I could use my compass when necessary, but I was counting steps and turns to keep my mind engaged and alert. I rarely depended on someone else to keep track of where we were. I was always on the lookout for tripwires and bamboo booby traps, too. We marched six feet apart, but that was still not enough room to prevent injury if a tripwire went off. This was only a problem when you were the lead. For the times that we suspected tripwires, we would send a mule ahead of us. We lost one mule that way. We usually marched in silence, so I had time to think and problem-solve. I had to remind myself to stay alert and not get caught up in too much thinking and daydreaming.

"When we moved into the bivouac area, we heard that the man killed in action was Bob Landis. All the men were emotionally drained and silent because Bob was well-liked from the first days that he boarded the *Lurline*. He was a very capable, experienced soldier who had fought in the Pacific as an infantryman—specifically, New Georgia. He was the lead scout for Blue Combat Team's I & R platoon under 1st Lieutenant William C. Grissom. We all thought it would have been one of us before him. We realized how vulnerable we were."

February 27, [1944]. Up and at em at 6:30 a.m. Move out. I led 4th dispersal group at Tawang. Cut my finger bad. Ate supper very hungry. Rations tomorrow. Hope I hear from the little woman. On watch at night. Fired BAR at nips heard them in brush. Slept well.

"I was on night watch that no one liked. There was a lot of responsibility not to shoot one of your own men while determining if the noise was from the enemy in case of a night attack that the Japanese loved to pull out of their bag of tricks. They also liked to attack at dawn. The Japanese were shrewd characters. They would hit bamboo sticks together or pop firecrackers to make more noise to intimidate us, so we would think we were outnumbered. So much of combat is patiently waiting, and then all hell breaks out, and you do the job you were trained for. When relieved of guard duty, I slept well."

"How did you cut your finger?"

"I was messing around with my Gurka knife. Most of us called it a Gurka blade or knife, but it was also referred to as a kukri knife. When I was at a Merrill's Marauders reunion, one of the guys said that there was a kukri knife that was owned by the King of Gorkha at a museum in Kathmandu. The kukri is one of the oldest knives, dating back 2000 years.[18] It's known for its razor-sharpness. When I was in a foxhole, I was trying to get a bug to crawl on the edge of the blade to see if it would cut the bug. That was my entertainment while I was waiting. Stupid, huh? I heard a noise, hurried to sheath the blade, and it cut my finger."

"Dad, you are so honest. So that's what you did while you were patiently waiting?" I laughed.

"Yeah. If I were in a foxhole, I would watch the wildlife or watch ants carrying a wing of some sort, or watch a bug crawl down my arm, stupid stuff like that. But I was always listening and in a ready position."

"Once, I saw a column of ants crossing a puddle of water. The first ant submerged itself, the second ant submerged itself on top of the first ant, and they continued until they made a bridge that the other ants crossed. Either they could hold their breath for a long time, or they were sacrificing themselves for the mission of getting over the puddle," I said.

"After I returned to the States, I wondered if our outfit was a sacrificial lamb," Dad said.

> *February 28, [1944]. Up 9 a.m. Cooked very frugal meal. Bouillon and crackers and plenty of coffee. Dried out pack and equipment. Had airdrop. Rations. Ate in a hurry then started out for Tawang Hka. Waded 5 rivers and 54 creeks. Left at 2 p.m. Bivouacked at swamp. Colder than hell. Too much noise and lights. Expect patrol activity. Slept cold as hell.*

"Third Battalion was in the lead when we crossed the Tawang Hka. Hka means river. Second Battalion followed us. We crisscrossed the river five times and crisscrossed the creeks 54 times. Often, the jungle would be so dense on one side of the river that we had to cross it for easier marching. The rivers were not deep because we were still in the dry season, but at night, it was cold and made us colder because of our wet clothes."

> *February 29, [1944]. Up 6 a.m. Built fires. We protected battalion. Colonel Beach left out in 4th dispersal. 500 to 800 Japs on our right flank. Enveloping move. Left out at 8:15 a.m. Crossed Chinese at least 9 more times. Sore as hell. Tired and wet. Marched 16*

damn miles. Arrived in bivouac area at 8 p.m. Cooked supper. After dark dangerous. Nip planes flew over. Bed 12 [midnight] Tired.

March 1, [1944]. Up 6:30 a.m. Cleaned up and rolled pack. Men had to shave. 4th platoon on detail for air dropping. Cooked only Bouillon, crackers, and hot pepper. Feet wet, will dry before moving out. Surprise no dropping. Built up stacks of wood for signal fires. But had to leave out at 7:30 p.m. Marched all night thru black jungle trails. Tired and sore as hell. Miss my woman.

"We were thankful for the morning mist in the jungle because the smoke from our campfires would not be detected. I was very tired and hungry, as all the men were, but we kept the morale up with our jokes and laughing. Later, when we got sick, was when our morale began to falter, yet we pushed those thoughts out of our minds. It was the only way to survive."

March 2, [1944] After marching in dark crossing 5 more rivers. 8:20 a.m. 12 hours of straight march. Kinda rough. Ate breakfast. The machine guns and mortars raising hell Just a short distance away. Slept for 1 hour in 48 hours. Awake and on move. Moved out after supply drop at 5 pm thru 6 miles of Jap positions. Bedded down and Lieutenant moved us to another place. The jerk.

"We marched for twelve hours straight in the dark. The men had their compasses attached to their packs so the man behind them could follow the light. Others that didn't have compasses placed a type of woody undergrowth that glowed in the dark on the packs of the guy in front of them. A certain type of fungi can be luminescent. It was slow going. I was glad I wasn't a muleskinner trying to keep the mules on the trail. I thought of Red and wondered how the guy behind her could see to follow. I guess they could hear her hooves plodding the trail. At one point, the guy in front of me turned and said that there was a gap in the line. Pass it on. I turned and told McBride, who was behind me, about the gap in the line and to pass it on. Soon it became, 'There's a Jap in the line.' In a short while, it came back to us. 'Throw him out.' We got a chuckle out of that one. For the most part, the men began to get angry over the night march. Falling asleep on breaks. Tired as hell. We finally stopped and slept on the trail."

> *March 3, [1944]. Up 5:45 a.m. Got pack. I & R 1 [p.m.]. Go into motion moving to trail block. For 20 min. Japs hit us. 2 men were wounded by shrapnel. Attacked again at 6:20 [a.m.] this morn. Moved out as rear point for column at 8:20 a.m. got within a mile and a half of our objective and we hit the Nips. 7 killed. I don't know how many were wounded. Moved on and bivouacked. Sleepless night.*

"Most of the fighting was done on the trails. We moved off the trails if they were firing at us, but most of the time, the jungle floor was so thick with thatch that you would sink into it, making it difficult to run. Our orders were to move to Walawbum, where the Numpyek Hka converged with the Kamaing Road. We established a block with our mortars and machine guns. We were to go through Lagang Ga while the 2nd Battalion was to go to Wesu Ga across the Numpyek Hka and establish a block further down the Kamaing Road. All was set, but we pushed so hard to make haste to get to Walawbum to cut the Kamaing that we didn't get to stay for the distribution of rations.

"In the meantime, our Orange Headquarters were in hip-high grass adjacent to a cleared field near a few raised bamboo structures of the village of Lagang Ga, where they encountered five Japs carrying a casualty about fifty yards away. They were finally identified as the enemy carrying a litter. The lead Nip saw our men and raised his machine gun, and our guys let the lead fly. All five of the Japs were killed."[19]

Later that evening, Mattlock and I were sitting around chewing the fat.

"Did you hear about Beck?" Matt asked.

"Charlie?" I asked.

"Yeah."

"I heard that there was some sort of commotion but didn't know exactly what happened. A wild elephant ran through the camp."

"Yeah. Charlie was leaning against a tree with a cup of coffee in his hands when a panicked elephant came crashing down the trail with a gash on its head. It was trumpeting a deafening scream. With a chain still around its leg, it ran straight toward him. Charlie froze on the spot. The elephant turned off sharply to the right and just missed him. It spooked the mules, and they took off. It took the muleskinners forever to find them. Funny thing was, Charlie told me later that as a kid, he was at a circus of some sort, and a runaway elephant came straight toward him and nearly trampled him."[20]

"Wow, what a coincidence," I said. "There's got to be a

message there somewhere."

"Yeah, Charlie needs to stay away from elephants," Mattlock deadpanned.

"Or maybe it was a way for him to face his fear and become more confident," I said.

"Stephenson, you're always thinking, my man."

"Well, we aren't exactly at a turkey shoot. Gotta keep the old brain active."

March 4, [1944]. Up at 6:00 a.m. Lots of firing all night. Started to move out as point. Fighting all around only God can bring us through. Moved out to relieve I & R. Snipers just about got our squad. I was the point. Dug in across river from Japs. Had a hell of a fight. 2 men got killed. Had ration drop but no food all day. Drank 3 cups of coffee. God saved us. Thank the Lord. Always trust and he will help. Expect heavy Jap fire tonight. B battalion coming up road with tanks, [The Marauders did not have heavy equipment, but the Chinese did.] A battalion in reserve. Chinese on the way. God, I hope mail comes in. I miss my darling. Would love to see her again. Please God send me back to her. She needs me. Please God.

We sat quietly for a moment.

"Dad," I asked softly, honoring the depth of the words he had just read, "What does it mean to be the point?"

"It means you are the point of the spearhead; you go first with the weight of responsibility for guiding the men. Sometimes, it is good because the Japs would let a few get by to draw a greater number of men to come out in the open, but most of the time, you are in real danger. It is one of the most dangerous positions you can be in."

"What does it feel like to be under fire?"

"As they say, there is no atheist in a foxhole. War is bad business. For the first time in my life, God was in the back of my mind every minute, a prayer always beneath my breath. I have never felt such desperation to live. The heart rate and blood pressure go up; your body is flooded with adrenaline. You feel exhilarated. It is the closest we will ever get to our primitive ancestors. When you are in the chaos of your life being threatened, it is hard to think clearly, but you learn to keep your mind calm. I would take a deep breath and pray stuff like, 'Oh God, keep my hands steady, my eyes and ears open, body swift, intuition right on, make me invisible to the Japs' or I would say something like, 'Keep us safe. Stay with me, Lord, I trust you, Lord, Amen.' Other times I felt like I was in a vacuum where my heartbeat was louder than the artillery. After a while, I learned to turn the volume down. I began to feel that nothing could shake my resolve. I became so engaged in the action that I no longer thought about myself, but I looked out for my family of brothers who were as close to me as my biological brothers. When I relaxed from the tension and kept myself and my muscles loose, I was a better shot. I felt like a true warrior."

> *March 5, [1944]. Still across river from nips. Small battle last nite. Col Beach led patrol across and just made it before all hell cut loose. Threw hand grenades. Still no rations. Fixed pack. Not much sleep. Dirty. God*

take care of us. No chow for 3 days but still going. Dive bombers opened up on Nips with bombs & shell fire. My squad & I went on patrol into Jap lines 4 miles. A man hit by sniper. Very lucky. God was w/ us. Slept well.

"After taking a pounding of mortar fire, 2nd Battalion withdrew its roadblock, which left only us, the Orange Combat Team, to retain control of the block at the Kamaing Road. Japanese patrols probed, trying to find our flanks, but we had already established ambush positions on the east bank of the Numpyek River. The Kamaing road was the north-to-south artery that connected northern Burma with southern Burma. It was a crucial supply route for supplies as well as new recruits for the Japanese. It was their lifeline that we had to cut off. At night, we heard their trucks, probably carrying supplies or new recruits. We heard the tailgates bang down. The mortar guys directed the 81mm artillery toward the road, and one round fell in the back of a truck of Japs, according to Sgt. Andrew Pung, who was observing from thirty feet up a tree."[21]

March 6, [1944]. Up 620 [a.m.] Shelling by mortar and mg all nite. Getting used to it. 4th pl moved out. 9 a.m. to protect airdrop. Back again at 12 noon. Jap mortars were shelling position very close. Finally got some rations. Not bad. Mahmood and Gorretti went to front. Dive bombers strafed and bombed enemy positions. Volunteers to get water. 530 [p.m.] Nips attacked for 1 hour. It was hell. They ran at us by waves, but our mg cut them down by the hundreds. Bullets cut the ground and leaves all around us. Artillery mortar, machine guns snipers just about got us out of ammo. Layed awake until 330 [a.m]. Sneaked out of positions and withdrew. 3 Letters to mail. 3 hours sleep. Only God could have gotten us thru.

"The Japanese withdrew their main force by continuous mortar fire and medium artillery on Orange Combat Team, which took the brunt of the attack; however, our men under Major Lew were prepared in deeply dug fox holes with logs placed on top. A mortar concussion near you shakes your whole being. It feels like it changes your heartbeat like it changes the rhythm of the universe. It is the vibration that shakes the interior. It feels so unnatural. Around 5:30 p.m., we patiently held our fire while two Japanese companies crossed to the west bank of the Numpyek River. Then we opened up with heavy machine guns, raining 5000 rounds of lead into the oncoming Japanese. They kept charging with the commands from their officers, 'Susume!' which means advance, and 'Banzai', their battle cry. Their bodies, arms, and legs were flung in slow motion across the battlefield as if in a macabre dance with the great equalizer—death. Some of their arms and legs were detached from their bodies. When the attack dwindled, 400 dead Japanese were scattered on the banks of the river. All of those men were brainwashed into believing that they were already dead. Now they were dead—never to feel the touch of a woman, never to feel their children in their arms, never to feel again.

"The Third Battalion was made of men who had fought at Guadalcanal, New Georgia, and New Guinea. They saw the atrocities perpetrated on our men, from mutilation to beheadings, and the desecration and cannibalism of the American dead."

I gasped.

"Yes, cannibalism! The men had evidence, like a slice of an American soldier's calf muscle found in a Jap's mess kit. It still had blond hair on it. The muscle matched the mutilated body.[22] These acts made the American soldiers from these campaigns angry, which boiled over and turned into hate. They loathed the Japanese enemy. These guys not only didn't hesitate to pull the triggers of their weapons, but they took retaliative pleasure in it. They knew they were fighting evil at its worst. Under this type of pressure, our men of the Third Battalion were hugely successful, not only as a well-oiled machine but as individual soldiers who learned that their weapons were their best friends. We became so familiar with our weapons that we could load and fire in our sleep. With that said, don't drive an American soldier to hate because they will release a holy hell on you."

I was shocked and blurted out, "Dad, that makes me sick to my stomach. Cannibalism? You've got to be kidding me!"

"I know," he agreed. "It was horrific. That is a part of the war that most of the population does not know about, and those who judge our soldiers as hard and crass have never been in a war. After you have fought in a war, then you can judge us. With the barrage of sights and sounds, you become desensitized to the point that you wonder if you were human. During the Jewish Holocaust in Germany, the inmates in the concentration camps became indifferent to the atrocities that were perpetrated on their fellow inmates. The difference was that they felt powerless to do anything about it because they would have been tortured or killed. We were empowered to stop the Japanese because of our equipment and guns. The powers that be knew that if you wanted to rend people to powerlessness psychologically and physically, take away their weapons. They will be at your mercy."

Dad's speech trailed off with the memories as he stared out of the long windows of the TV room into the green, yellow, and blue of the landscape and sky.

"Was it worth it?" he asked quietly. "I have regretted my rash act of volunteering. I wished it had not been me stepping forward in all my young bravado to fight over there. But if not for me, then who? Fighting over there instead of my own backyard helped me to make that decision. The men I fought with were my brothers—my family. When we were in dangerous situations, I trusted my men to have my back, and I protected them as well. I have often thought of those men and the ones who survived. I wondered about their lives. Most likely, they buried their memories just as I did until now. With all that said, I was proud to be a Marauder and to use the skills that the United States Army taught me. I would not hesitate to do it again. I would fight for my country. I never took a pension because I felt that I was only doing my duty. My point in my regret was that it has been hard on the family. I know I am a workaholic, and I stay too busy so that I can forget, but sometimes I am not present either."

I sat clueless, not knowing what to say because he was right. I couldn't speak because of the lump in my throat.

> *March 7, 1944. Hit the deck 7 a.m. Mortar and machine gun fire still on. Moved on to trail while air drop [___?___] then drew rations and moved out at 12 noon. Man was killed at air drop and buried. Moved 3 miles. Arrived 5 p.m. Cooked up chow. Found rice and corned beef. Tasted swell. Bed 8 p.m. Saw old buddies and Chambers good boys. God pulled us through.*

"I later learned that Pfc Carter Pietsch was the soldier killed when a parachute failed to open for an airdrop of ammunition, and it dropped unexpectedly near the aid station. So sad," Dad said as he shook his head and paused. Walawbum had fallen into American hands on this day, March 7, 1944. Of the 2,750 Marauders who fought at Walawbum, 2,500 remained to fight. 800 Japanese were killed. According to Dr. James Hopkins, 3rd Battalion surgeon, 'the victory at Walawbum had allowed the Chinese to advance more in five days than they had been able to advance in five months.'[23] We later learned that the Marauders, along with the Kachin guerillas under Lt. James Tilly, and the Chinese were the only allied troops fighting in Burma at the time of the Battle of Walawbum.[24.]

"Although General Stilwell never met the troops, he sent his congratulations for the rigor and determination the men exhibited when taking Walawbum, which was a huge morale boost for all the battalions—or at least the ones that heard it. The happy-go-lucky attitude of the Orange Combat Team was contagious. The men of the Third Battalion had an unorthodox battle cry that was like a wolf howl, which made us sound like we were not human. The Japs already considered us ghost soldiers because we cleaned up our bivouac areas and buried our dead. Our battle cry was actually funny because we were yelling at the Japs—ASSHOLE and drawing out the 'hole' in a long wolf sound.[25]

"We had a great group of guys. Some were street savvy, and others were good old country boys who could shoot the dot out of an i. They were so funny and jubilant when we got the best of the Japs. It was like a football game when we scored. When both sides were out of ammo, someone would stand up and yell obscenities like, 'Come and get some more, you yellow bastards!' An English-speaking Jap yelled, 'Americans eat Spam!' Another yelled, 'Eleanor Roosevelt eats powdered eggs!' and finally, one of the English-speaking Japs yelled, 'Babe Ruth stinks!'[26] At one point,

Major Petito ran right out in the middle of the battle and yelled, 'Tojo eats shit!' Whoa, that one got the guys going. We later laughed about the whole scene. Humor kept us loose and relieved tension, but if you heard the tales of fighting in other areas of the world from these guys from Guadalcanal and other islands, you could feel anger simmering beneath their exterior."

"So, laughing obviously helped relieve stress. Any other ways of reducing stress from all the firefights you were in?"

"Yeah, we laughed a lot, but we never underestimated the danger that we were in. I would take a deep breath and hold it, then exhale. This helped to calm the nerves. Also, breathing deeply and bringing my shoulders forward and back helped calm me down and loosen the muscles," Dad demonstrated. "You get used to the noise and chaos," Dad continued, "Sometimes it felt great to lose myself in the moment. That happened when I didn't overthink things and just methodically fired."

> *March 8, [1944]. Up 7 a.m. Nice & sunny. Drew rations went over and saw the pictures that Ethel sent were swell. Good [_?_] Rice and bouillon. 1st bath after 2 weeks. Rested pulled out at 12 noon. Marched 7 miles. Arrived at Bivouac area at 820 [p.m.] Bedded down. Contacted 10-man patrol.*
>
> *March 9, [1944]. Hit the deck 7 a.m. moved 300 yards to river trail. Set up defensive positions. Ate 4 times Rice and other things. Sick as hell. Took bath and cleaned-up. Have dysentery. Sick as hell. Lt. Weingartner is new platoon leader. Lt. Burk was reld.*

"I remember being sick with dysentery, as well as many of the other guys. It was so bad that we cut a back flap out of our fatigues for convenience. The Chinese bivouacked with us. They were Sun Li Jen's 113th Regiment of his 38th Division. We were grateful that our allies were with us to help in the fight, but there was a clash of cultures. They would take anything they could find. So, you had to carry your packs and belongings around them. According to Theater Policy, we were not to consider it stealing. We had to be watchful and tolerant of this behavior. The one behavior that we learned too late was that the Chinese used the rivers for latrines. Unlike the Chinese who boiled their water, we used halazone pills that we put in our canteens, but they did not kill the bacteria that caused amoebic dysentery."[27]

"I know it is made from a chlorine compound, but it should have been tested before taking it into combat," I said.

"Who knows? Maybe it was tested. Science has made great strides since then. It would have saved a lot of misery for sure if we had known that." Dad nodded.

"You've got that right," I huffed. I had recently found a 1991 report by the National Toxicology Program Executive Summary of Safety and Toxicity Information on Halazone. It stated that halazone in humans caused 'dermatitis, rhinitis, conjunctivitis, and bronchitis as well as asthmatic symptoms.'[28] With that being said, I suppose it didn't help when you got Typhus and coughed up blood.

March 10, [1944]. Up early 620 [a.m.] Everything alright so far. Sicker than hell. Stomach weak. Also, ate chow twice. Was issued ratns for two days, Heard news broadcast publicized Yanks in Burma. Had air drop. General and troopers went to Mangwa [Maingkwan.] Bombed for year and ½. Not much else.

I was glad that the American public knew that we were in

Burma. I thought maybe Evelyn would hear the news and know where I was. Oh, how I missed her."

March 11, [1944]. Up this morn. Weak as hell. Chest and kidneys hurt. Got airdrop 10-1. [_?_] I Ate quite a bit. Layed around. Took bath and washed clothes. Got clothing issue. We are to be on the left flank moving out tomorrow at 10 a.m. Still feel like hell. God, please give me strength to go on. Slept under parachute.

"It was great to rest. When my body faltered, I worried that I would not be able to keep up and would be left behind. I didn't want to be separated from my platoon. It was far too dangerous."

March 12, [1944]. Hit the deck 6 a.m. Policed up area. Dried out clothes. Fixed pack. 4 days rations. Big load. A BN came by at 9 a.m. Saw all the old boys. Good to see them. Still weak as hell. Hope I can make it. Moved out at 1030 [a.m.] Behind 3rd platoon. Marched 16 miles. Rough as hell. Crossed river and arrived at LaJauge. 5:45 p.m. from Ninglap Ga. Ate breakfast food, coffee and went to bed on air mattress. Chinese fired into bush and killed cows. Feel better.

March 13, [1944]. Hit the deck 630 [a.m.] Dried out and made pack. Ate soybean cereal, Will move out at 9 a.m. Have 6 days rat because I couldn't eat. The Burma Raiders will march 12 miles today to Kadau Ga. Cross 5 rivers. Followed to Pabum. Boy it was rough. In at 445 [p.m.] Marched 31 miles. Rough as hell. Tired and sore. Passed last Chinese out post. Took cold bath in Tana [Tanga] Ga. Ate supper, to bed at 8 p.m.

"When you are in survival mode, your body can take a lot of stress because you have no other choice. Psychologically, you never let your mind feel sorry for yourself. You never think 'Poor me.'"

March 14, [1944]. Hit the deck 630 [a.m.] Still have 4 days rations. Very heavy. Packed up. Slept swell. Ready to move out. Moved out at 945 [a.m.] went 8 miles. Kinda slow and tiresome. Into bivouac area at 3 p.m. Built fire and dried out clothes. Bed early. Very sore.

"Thankfully, I got some rest that helped me to heal. I felt out of sync with the rhythm of the universe, but nothing could stop my prayers. I prayed for healing and guidance, and I prayed to be happy."

March 15, [1944]. Hit the deck early. Men were issued shoes and rations [for] 3 days. Moved out as point. B Bn is in front of us. Khaki in rear. Marched 7 miles. A lot up hill. Rough. Nice Bivouac area.

After we set up camp, I went by the mules and horses to check on Red and to say hello to J.D.

"Hey, girl," I said as I gave her a sugar cube. I brushed her with my hands and hay until you could see a little glimmer of a shine. She was so skinny. I prayed that the planes would be able to drop additional hay and grain. It felt good to think of her problems rather than my own.

"We're hanging in there, aren't we old girl?"

> *March 16, [1944]. Up and at em 730 [a.m.]. Left out as rear guard. Passed B Bn 6 miles. Rgt HQ Had very bad hills. Arrived at bivouac area 0820 [a.m.] Marched 16 roughest hills and miles I have ever been on. Rained like hell. Got wet. Bivouacked after eating. Bed late. Had guard.*
>
> *March 17, [1944]. Up 5 a.m. Packed and left out 7 a.m. Marched 4 miles for air drop. Johnny Kazoski [Zokoski] was killed. 5 holes in him. Saw doc work on him for 2 hours. Dried out clothes and blanket. Took Bath. Was issued rations & K & C & candy. Swell. Ate good chow. Had a hell of a stomachache. Built nice shelter with banana leaves. Smuck and I. Rained like hell. Kept dry.*

"This was a sad situation because it was an accident when Johnny was cleaning his gun. A round was left in the chamber by accident, and the gun blew up in his hand. Multiple shards of metal went through him. When he succumbed during emergency surgery, we were all overcome with emotion. Lieutenant Weston performed a burial service, and we buried him right near the accident site," Dad said.

"Oh, Dad, I'm so sorry," I said.

A few years later, I met Johnny Zokoski's sister at a Merrill's Marauder's yearly reunion. She was desperate to know about her brother Johnny. After all those years, she was haunted and was begging for information from anyone. Someone told her about my dad's diary, and we spoke. When I got home, I looked in the diary, and there it was. The entry above. I sent a copy of the entry to her. She later sent a card thanking me and stating that she was grateful to know the effort the doctor made to save her brother's life."

> *March 18, [1944]. Up early 630 [a.m.]. Fixed pack. Found out we are staying here. Ate meat and bean stew. Darn good. God is kind. Layed around. Not much doing. Wish God would send some mail. Have been getting a good appetite. Bed 830 [p.m.]. We will be out of here in 20 days. Got parachutes to sleep on.*

"According to the promises that were made to us when we signed up, we thought we would be out of there in 90 days. January through March made 91 days, including the extra day in February for the leap year. Rumors were swirling around that we would be out of there in 20 days. I didn't know if it was true or not, but it gave us light at the end of the tunnel."

> *March 19, [1944]. Up 7 a.m. Had nice sleep. 4th plt is on airdrop detail. Drew 3 days rations for 2 days. Expect mail today if God is kind. Ate C rtns, got water, shaved and aired out my feet. Miss you honey. Had trouble with McBride. Went to river with old dad Goddard. Washed shorts and socks. Took bath. News, we have completed one more mission. A Battalion lost few men. We have one tough battle ahead. Ate stew, cheese, bouillon. Rice [and] coffee. Feel fair. Move out in morning. Bed Early.*

> *March 20, [1944]. Up at 715 [a.m.] rolled big pack. 4 days rations and ready to move out at 8 a.m. Finally moved out 11 a.m. Marched 3 miles all up hill. Was Roughest yet. Just barely made it. Arrived at Janpan area on Road block trail leading NW. Fixed shack. Have 2 BAR's on trail. 3 hour guard 12-3 a.m. Bed early.*

March 21, [1944]. Up and at em 0600 [a.m.]. Woke up Mahmood. Last shift. Cooked crackers, bouillon, pork & egg yolk & cheese. Tasted swell washed down with coffee. 2nd Squad takes over trail block 12 noon. We're at Janpan. 25 miles from Shaduzup. A BN is there. 2nd BN & Khaki column is at Warong. 25 miles. We will get airdrop of 10-1 rations today. 3 p.m. got chute and 3 extra rations then 1 box of half 10-1. Ate till I could not eat more. Fixed pack and bed. Couldn't sleep.

"General Merrill, who had his command post at Janpan, gave orders for the Orange Combat Team to move quickly to block the southern trails around Auche, Warong, and Manpin," Dad explained.

March 22, [1944]. Up 545 [a.m.]. Made pack with 4 days of c rtns. Extra candy and stuff. Ate good old 10 &1 for breakfast. Pulled out at 8 [a.m.] behind 3rd plt & BN then us. Have 9 miles to go. Darn rough marching up big hills and down. Pulled in at bivouac area at 3 p.m. & put up trail block leading to Warong and Kamaing. Rained like hell. Gosh got letters & more letters from Evelyn 24 letters. Frankie 4 letters and Paul 1 and Mr. and Mrs. Hower 1 letter. 32 in all. Good news from CBI on Merrill's Marauders. Big deals.

"Out of all of Evelyn's letters, one stood out and reassured me of her love. I treasured it. It kept me alive. After reading it, I folded it tightly and slid it into the cellophane sleeve that I kept my diary in and put it back in my shirt pocket."

Dear Larry,

The newspapers are writing about Merrill's Marauders in Southeast Asia. I feel in my heart that you are there. My darling, I know you are in trouble. Don't ask me how I know, I just do. Love can open you up to things of the mind and heart that are not bound by the laws of this earth. I am here to tell you that you are coming home to me. I am calling on the Army of the Most High to save you, and you will be saved. I have babies in me that I need to give birth to and you, my love, will be the father. God will not let us down because he has a destiny and purpose for us and our unborn children. You will come home to me.

Love,
Evelyn

March 23, [1944]. Hit the deck and moved out towards Warong at 7 a.m. Rough as hell up hill and down. Had to detour around Warong. Turned off trail to right, covered 14 miles. Bivouacked after crossing 8 rivers. Nice fire, food, and sleep. Native guides slept near us.

March 24, [1944]. Again hit the deck and started out at 730 [a.m.] as rear guard. We went 9 miles & put up rear trail block near small village. Expect air drop tomorrow. Waded 16 rivers. Wet & tired pulled in at 1130 [a.m.] Built fires and ate good old c rations stew. Don't know what is next. But God will guide and protect us. In valley now near Kamaing. 4th plt moved out to Poakum 5 miles. Stayed with I & R. Nice. Rest. 1 [hour] 15 min guard.

"My friend Lieutenant Logan Weston was verbally ordered to proceed south through the towns of Hsamshingyang, Nhpum Ga, and Auche to the small town of Manpin, where he was to establish a trail block. Weston's Platoon scouted the trail twelve hours in advance of the approach of his parent 3rd Battalion. He established his perimeter-defense type block upon reaching Manpin at dusk on March 24. He was ordered to hold there until the battalion arrived. The 2nd and elements of the 3rd Battalions continued westward through Manpin to Inkangahtawng and cut the enemy's main supply route, leading to Japanese troops fighting the Chinese farther north.

"Upon arrival at Manpin, Weston assessed the terrain and the location of the trails. He radioed back to his battalion and recommended that he be authorized to proceed southward about four miles toward Kamaing to Poakum and establish his block in the vicinity of Poakum. There was a known reinforced enemy battalion and some support forces at Kamaing, located twelve miles to the south. The Battalion Commander, Colonel Beach, granted approval, and Weston moved his platoon at first light on March 25, 1944, to Poakum.[29] There was speculation that the enemy in Kamaing might take the trail through Poakum to Manpin to cut off this single escape route. Lieutenant Weston's strategy was to draw the flanking enemy forces up the south parallel trail toward Warong so that the north trail to Auche would be open as a route of withdrawal for the 2nd and 3rd Battalions after accomplishing the roadblock mission in Inkangahtawng.

"Lieutenant Weston established his platoon-sized perimeter block at Poakum, then sent an ambush patrol 3.25 miles southward toward Kamaing. This patrol ambushed and killed a twelve-man enemy patrol coming up from Kamaing a few hours later. Our 3rd Battalion's I & R Squad was ordered back to Poakum, where they joined the platoon perimeter defense. A few hours later, strong enemy forces came up from Kamaing and were assumed to be the advance guard of a company. It later turned out to be the advance guard of a battalion!"

"Wow," I exclaimed, "Sometimes ignorance is bliss. He didn't know he would have to face a whole battalion."

"The Japanese had 1,100 soldiers in their battalions," Dad said. "This enemy force attempted to encircle Lt. Weston's platoon by enveloping both to the right and left. When it appeared that the I & R Platoon was about to be surrounded, Weston displaced eastward on the south parallel trail to the next best defensible area and established a second block. When that block threatened to be encircled or eliminated, he again withdrew to a more secure location toward Warong. It took the Japanese three days to drive

Weston's platoon four miles through the successive trail-blocking positions. It worked. The mission was being accomplished because the enemy pursued Weston up the south parallel trail, keeping the trails from Poakum to Manpin and from Inkangahtawng to Kauri open as the escape route for the 2nd and 3rd Battalions.

"The second day of the withdrawal action toward Warong, an enemy knee mortar made a direct hit on the I and R's radio, knocking out the communication between Lt. Weston and his battalion commander, now located near Manpin. On the third day of this action, the battalion commander sent our platoon under the command of Lt. Warren Smith to Poakum, then to Warong, fearing that the enemy may have eliminated or encircled Weston. When we arrived, Lt. Weston had been driven back toward Warong on the Poakum trail. We, Lieutenant Smith's platoon, secured Weston's east flank and then proceeded to establish a block on the Warong-Tatbum trail.

"On the fourth day, another Jap company, supported by mortar and artillery, came up from Tatbum and ran into our block at Warong. At the same time, Weston was being pursued by the enemy coming up from Poakum. He leapfrogged his unit by squads over 3.5 miles on the Poakum trail near the outskirts of Warong to avoid encirclement. Lieutenant Smith, fearing our platoon would be annihilated, leapfrogged us from Warong northward through Auche, Kauri, and Nhpum Ga.

"Dad, what is leapfrogging?"

"It's a military technique used to move the troops from a particular area when you have advancing enemy fire. You usually have two teams. We went by squads. One squad or team suppresses the firepower of the enemy by rapid-firing their ammo while the second squad is repositioning. In this case, they were backing up the trail. On a signal, they move, and the second team comes up and maintains fire against the enemy, and the first team repositions. This way, little by little, you can move troops away from the pursuing enemy without turning your backs on them. It is sometimes called 'peel.' It is used for situations when smaller groups of infantries like Weston's I and R disengage a much larger force and withdraw up the trail.

"By the time we were driven northward as far as Auche, the 2nd and 3rd Battalions then proceeded north four miles to Hsamshingyang. The 2nd Battalion dug in a battalion-sized perimeter at Nhpum Ga. At the end of the day, the enemy had driven Lieutenants Weston and Smith eight miles in four days. They patrolled the trail as far north as Hsamshingyang, where the 3rd Battalion was located. The 2nd Battalion was unable to move and established a stabilized defensive position. The enemy surrounded the 2nd Battalion at Nhpum Ga. An eleven-day battle ensued and tested the mettle of our platoons. Ninety of the enemy were killed by the action of Weston's and Smith's platoons during their blocking and withdrawal action. During the action, Weston's platoon lost two mules but no personnel. On two occasions, Weston received shrapnel wounds from enemy grenades and knee mortars, and on the third day, he suffered a recurrent attack of malaria. He refused to be evacuated because of the dire circumstances of the situation. Also, during this action, Weston ran out of ammunition for his personal M1 rifle on two occasions. He resupplied his unit of fire by re-distributing ammunition from the men of his platoon and our platoon led by Lt. Warren Smith. The ammunition was sufficient to keep the I & R platoon effective and avoid surrender or annihilation."

March 25, [1944]. Hit the deck at 600 [a.m.] Ate breakfast. Have 8 rough miles to go. All up hill. It's going to be rough as the devil. Marched 8 miles to Warong. Have trail block to Auche. The I & R was hit 2 times. Killed 9 or more Japs. No losses. We are in a bad spot. Mahmood went to Auche to take message. They sent rtns by mule to us. Just about out. God will take care and protect us. Ate mostly bouillion and coffee. Bed 7 p.m. Fresh Jap tracks.

March. 26, [1944]. Up 5-7 on guard. Kinda rough. Moved up hill to another trail block to Ashau & dug in. Thank you, God for the protection and guidance last night. Ate breakfast then Nips hit us with about 600 men. Boy it was rough. Total nips killed about 25-28. We pulled out. Fast withdrawal. Boy it was rough. Haven't washed in days and am tired & sore. Will make last stand at Auche. 3 miles from Warong. I pray to God to give us protection and carry us safely thru this campaign. God in heaven give us help. Please God send us food and ammo. B [2nd] Battalion came in at 10:30 a.m. C [3rd] Battalion came in at 3 p.m. Have a very good perimeter. Got 5 days of rations by air drop. No sleep again tonight. Nips all around us. Boy its rough as hell.

"Withdrawals are rough because you must move fast. You literally outrun them. Your adrenaline is at full throttle."

March 27, [1944]. Hit the deck 530 [a.m.] Rolled packs and started out at 6 a.m. I & R, 4th Platoon, and rest. Boy, we ran 9 miles. Artillery, mortar, machine gun and everything. I ran till I was out then they tied me on a horse and my pack got lost. Arrived in Sam Shing Yang

[Hsamshingyang] 3 p.m. The doc worked me up. Out. Only God can see me through. Please God in Heaven help me get safely thru this campaign and safely back for Evelyn's sake. McBride and I slept in fox hole. No quilts or nothing. Cold as the Devil. Swagger and Smith helped me. Thank you, Lord.

Dad sat quietly and thumbed the pages of the diary.

"March 27 and 28 were rough as the dickens. Raw days," he said, clasping his hands and shaking his head. "I ran until I passed out. I was out cold. Maybe it was low blood sugar since we didn't have any rations, or maybe it was poor nutrition or just being plain worn out. I know the doctors would diagnose a patient with AOE, which stands for 'Accumulation of Everything.' When you lose control like that, you are so vulnerable. Sometimes, I think that the medic that saved me was an angel sent by God."

"Mac said that he called for a medic, and one came and tied me to a horse. I liked to think that it was Red who got me out, but I never knew. The mules and horses were invaluable to our campaign. They suffered as much as we did during their service time. Often, they took a bullet or shrapnel and saved one of our guys.

"Mac and I both lost our packs. Our packs carried all sorts of things for our survival, especially our rations and halazone for purifying our water. There was nothing to cook with. It had my compass in case I were to get separated from the unit. I carried my map in my pocket. It carried a change of clothes and a blanket for cold nights. It carried my souvenirs, like Jap grenades and knives. There was nothing to protect my body. I knew how it felt to have nothing but the clothes on my back and the tags that identified who

I was. Our packs carried letters from home. We were cut to the bone when you have nothing but the clothes you are wearing. I lost my emotional lifeline when I lost the letters from Evelyn that I had read and reread. Her letters were the last tangible pieces of paper that her hands had touched and where she had poured out the sentiments of her heart with her pen. The gut-level survival was all I had. I was living the ragged sound of one bound to the ground and unable to escape.

"I wanted to cry, but tears would not come. I feared that I was too hardened and desensitized to ever cry again. I began to think of good things, positive things, and I realized that I had my will, I had my faith in God, I had my soul, I had my determination, and I had the memories of my wife, who I was going to see again. I had the angels surrounding me, and I was grateful. I remembered my diary and touched the pocket of my shirt to make sure it was still there. I pulled it out of the cellophane sleeve that it was in, and yes, my favorite letter from Evelyn was there. I had forgotten that I had put it there. Then I told Mac, 'We are going to get out of this, buddy. This is temporary. They didn't name us provisional for nothing. We are going to get out of this godforsaken place. Believe it, Mac.'"

Nhpum Ga

"The infamous siege of Nhpum Ga began on this day, March 28, 1944. General Merrill ordered Colonel McGee's Second Battalion to hold Nhpum Ga while Third Battalion seized Hsamshingyang four miles to the north. Hsamshingyang was a good location to build an airstrip for an evacuation point as well as a supply base. Nhpum Ga was a small village of a handful of basha huts on top of a mountain at an elevation, I would say, around 2,700 to 2,800 feet above sea level. It lay between two watersheds of the Tanai River on the east side and the Hkuma River on the west. It was accessed only by a narrow north/south trail along a steep and dangerous ridge of the mountain. The trails were surrounded by dense tropical undergrowth and tall trees with sheer 1,000-foot drop-offs. The canopies of the trees were covered with a woody type of vine that I later learned was called a liana vine. It created a tangle of webs that competed with the trees for sunlight."

March 28, [1944]. Up 5:45 a.m. Cold as the devil all night. Drank coffee for breakfast. Couldn't much more. Very weak. Helped in Air drop then 1st squad went upriver and formed ambush. Gorretti brought chicken and pie up. Moved back to positions at 530 [p.m.]. Fell asleep in parachutes. Mac and I both lost packs to Nips. Drew 10 and 1 rations and clothes.

"General Merrill had a heart attack on this day, and Colonel Charles Hunter, who was a West Point grad, assumed command. It was dark times. We admired General Merrill, but we also respected Colonel Hunter. No matter how down we got from our circumstances, we tried to stay positive."

> *March 29, [1944]. Up 530 [a.m.] 4th Platoon is going up trail to contact B Bat. Got trapped and stayed until 2 pm & then had to disperse and run down to river. Just made it by 5 p.m. God was with us to pull us out of a tight like that. Made smaller perimeter. 3rd Platoon moved up to secure trail. Rained like the devil.*

"The 3rd Battalion patrolled the northern trail to Nhpum Ga as well as the newly constructed airstrip at Hsamshingyang. The overpowering number of Nips trapped our squad, and we had to wait it out. We stayed quietly in the cover of the jungle and discussed our plan. We had to be patient and let the sun go down a little more so we could make a dash for it. I kept looking in the trees, trying to locate a sniper. We thought we had found him, so our plan was to pop a smoke grenade and throw a hand grenade in the opposite direction from where we were going to cross the river. It would kick up dirt and debris, throw some shrapnel around, and create a little chaos that would keep their attention away from us. We had to do anything that could give us cover. When the time was just right, I threw the grenade. We protected ourselves until one of the guys detonated the smoke grenade, and at last, we spread out and ran like hell toward the river. Talk about walk on water, I think we did—we skimmed right over it. They were shooting, but my entire squad made it without any injuries. It was amazing that we all survived.

"It was dark in the jungle when we made our way back to Hsamshingyang. When it rained, the trail became a muddy hell—slick as a mountain of ice. I have experienced rains in Louisiana and the Gulf South," Dad said, "but the rains in Burma were like silver spears streaking through the sky. It was a hard rain. I tried to trick my mind into thinking that I was taking a well-needed shower, but in the mountains, it was usually cold at night, and you ended up not able to sleep because your body shivers in wet clothes. Ponchos helped, but I had lost my pack with my poncho in it. I wanted to scream my frustration at being miserable!

"The rain did, at least, wash the hard, dried patches of blood made by leeches. Those pesky, rubber bloodsuckers gave us fits. They were on the vegetation and could get on us when we brushed against underbrush or the tall kunai grass. Those little suckers could hang on for the ride until they could find a nice, juicy piece of tender skin that they could dive into. They could become so engorged with blood that they would balloon up several times larger than their original size. When they clamped on, the best remedy was to use a lighted cigarette held close enough for them to feel the heat. They would let go, but the anticoagulant in their saliva made red rivers run down our legs or wherever they had a mind to attach. One guy got one in his nose while he was sleeping. Sometimes, I was so cold in the foxhole that I couldn't steady a cigarette to get the leeches off. I would give up and smoke the cigarette.

"The mules and horses had the same problem. Whenever I saw Red, I got the leeches off using my cigarette and putting pressure to staunch the blood flow. When I could, J.D., Red's handler, and I would plaster mud on her from her hooves past her fetlocks to ward off the leeches. Red was a trooper, calm in the face of danger and the constant barrage of noise; however, when she smelled a wild animal, she got nervous and was known to bolt. Red's smell and the leather pack saddles always took me back to the days of my boyhood when I went on trail rides in the mountains. It helped me to remember happier times and made me more determined to survive. I wanted to see Mama again."

Dad and I sat quietly for a minute. I bit my lip to control my tears; the mama sentence got to me. I didn't want Dad to stop for fear that I wouldn't be able to take it. Dad stared out the window. I felt an overwhelming sadness for the soldiers and him. He must have been very scared—his body failing from diseases, poor nutrition, and exhaustion physically and mentally. I thought of him in dire straits, thinking of his Mama. It must have been a comfort for him to think of the nurture, support, and love she gave him. So many of the dying called for their mothers. Their mothers were the first human beings who loved them and the first human beings desperate for them to survive. I remember when I had cancer and was seeing my bald head and my mouth moving from above as if I were floating around the ceiling. In an instant, I was back in my body and heard my voice screaming for Mama. I understood their desperate cries. I also realized that I had always regarded Dad as a hero. I realized how young and vulnerable he was and how afraid he must have been even though he hyped himself up to be a warrior.

March 30, [1944]. Moved up trail to Kauri to get B Battalion out. Started raining. Mg 31 Hit us 3 times. Just missed Mahmood & Taylor and Paul and I. Moved around flank 3 hours. Cole got hit and 4 other guys. Rough as hell. God, please guide and protect us & get us back to the States safely for Evelyn's sake in Jesus' name amen. Formed perimeter on trail at 5:30 p.m. Cold, wet. Raining. God was the only one that saved us. Didn't sleep, too cold.

"Cole was shot at about 20 yards. He was on the trail, covering for one of the other men. The guys got him out and to the back, where the medics carried him to our battalion surgeons. I said a quick prayer that Cole would be okay. He was carried by a litter and evacuated at Hsamshingyang airstrip. He recovered for a little over a month, then was sent back to battle."

> *March 31, [1944]. Up 530 [a.m.] Started up trail at 6. Ran into machine gun 31 again. Graham got it again in arm. Jack Pedro [John Ploederl] was killed. God, please guide and protect us in Jesus name. Amen. Big Robinson Robby [Edgar Robertson] was killed. 7 men were wounded and 1 killed by knee mortar. All were radio operators. Can't go forward. 2 75s were dpd to be used by us. P-40 strafed and bombed positions, but it didn't help. So far, the trail is clear. Cold as hell.*

"When General Merrill was evacuated to the hospital in Ledo, he gave orders to have two 75MM Howitzers dropped to the 3rd Battalion at the Hsamshingyang airstrip. What a boost in morale they were. Sergeant Acker of Kahki Combat Team was experienced with this weapon and could put it together without the instructions. When it was brought down the trail, the men stepped out of the way and cheered.

"The big dog's off the porch," Taylor said. "Finally got some firepower that'll bite ass."

"Later that evening, when everyone was so down, Mattlock came by to check on me. 'We're gonna win this war and make it home,' he said to me, echoing my earlier words. The guys and I huddled together in our grief at the loss of so many of our friends. Mattlock said, 'We gotta keep going. We can't let our friends die in vain. We are going to win this campaign. We are going back to the States to let everyone know the sacrifices that we have made for our country and its citizens.' I will always remember that uplifting, positive attitude from my brother-in-arms. This was a day of huge loss to our platoon and the people killed or wounded, but a day that we made offensive gains. We fought through two trail blocks and gained a couple of miles. I remember when there was a lull in the fighting, I couldn't help looking at all the brave men and wondering if they or I would be alive the next time we got a moment's break. We were all running on adrenaline and fighting for our lives. I remember something that my mother had told me when I was a little boy. Teach a child to be self-confident with a goodly conscience, and you raise a hero. Be a hero in your own life and especially for the lives of others, and always remember 'mind over matter.'"

April 1, [1944]. This may be April Fool's Day, but there is no fooling around today. A bn patrolled right flank. Contacted nips. There [they're] all dug in. Khaki came thru with bed rolls. No chow as yet. Still cold. Had set ambush on waterpoint. Was relieved. God in heaven please guide and protect us and bring us safely through this & take me back safely to Evelyn for her sake. In Jesus name amen. Got blanket and poncho. Rice was sent up by Khaki. Tasted swell.

April 2, [1944]. Up 5 a.m. Expect Nip attack. Cold as the devil. Cooked cup of tea. Swell. Some more rice from Khaki. Still trapped. God brought us through the night safely. Expect they break out today with God's help. On water ambush. Will attack at 230 [p.m]. 4th went around left flank for diversion. Expect dive bombers & Art blast loose at them. We fired and got the hell out. 1st squad led. The Nips followed & fired on us. Bivouacked on hill. Cold as hell. God is with us.

"I missed recording in my diary on April 3rd and April 4th. Dr. James Hopkin's book *Spearhead* reminded me of the events of that day. At dawn on April 3, Orange Combat Team probed the thick jungle to learn where the Japanese positions were. We were not to engage in offensive action.[30] Most of us stayed near the trail. April 4th brought a new game plan. Orange Combat Team was to go up the side of the mountain for about 200 hundred yards. The Japanese were embedded in bunkers. The planes bombed and strafed the Japanese positions, and we laid a barrage from our 'fat boys' artillery. We charged the Japs before they recovered from the shock of the artillery onslaught."[31]

> *April 5, [1944]. Somehow, I missed 2 days. The Lt. led us in a circle trying to get out. Then Barney led us on the trail. Major Lou [Lew] was wounded. Captain Burch got Jap mg. Tired and sore. Artillery and planes raising hell. God be with us in the coming fights. Saw Mosier, swell guy. Got me water. Went up and built-up perimeter around 75 and 81 mortars then built-up perimeter at foot of hill next to Nips. Didn't sleep very much.*

"Major Lew was hit in the lower chest on this day. A couple of guys and I crawled to him and pulled him off the trail. We got him to the rear, where the medics took over. This was a huge blow to our company. The surgeon gave him blood and patched him up. He was evacuated to the hospital in Assam, India, the next morning. I heard later that he recovered from his wounds."

> *April 6, [1944]. Up 5 a.m. Rolled pack. Big push today. Boy, I mean push. The artillery and mortars and dive bombers laid down a big barrage that [was] very close to us. Deafening then all was silent, and we moved around right flank to protect Lt. Weingartner's platoon.*

Moving up trail thru Jap positions. Lots of dead laying around where our dive bombers got them. Saw them rush thru woods. We protected trail to rear. Gained ¾ mile. 1st squad went up to feel out Japs on hill. They cut loose with machine guns and rifles. It was rough getting out of there. Bullets all around us. Tayler got burned. 1 day rat. Dug in. Mc and I had tea, cheese, pork, and eggs. Our Artillery shelled Nips.

"I later found out about the heroism of our Nisei interpreters, especially Sergeant Roy Matsumoto, on April 6, 1944. The Nisei, whose parents were Japanese immigrants, were born in the United States and served as G-2 men in Army Intelligence. They were considered second-generation Americans. There were a couple of places in the United States where these Nisei trained to be interpreters, translators, and interrogators: the first school was the Presidio in San Francisco, established in 1941, and graduated the first class who were promptly sent to Guadalcanal. The other was at Camp Savage, Minnesota, which housed the Military Intelligence Service Language School (MISLS).[32] Those assigned to Merrill's Marauders were Roy Matsumoto, Ben Sugata, Grant Hirabayashi, Jimmy Yamaguchi, Henry Gosho, Calvin Kubota and others. I think fourteen in all. The value of secrecy for these men was necessary. If they had been seized by the Japanese, they would have been tortured to death. Their relatives in Japan would have been singled out and killed. When Matsumoto left for San Francisco by train, the windows of the car he was in were curtained.

"They were invaluable to the Marauders for their bravery – they would crawl out to outposts in order to overhear the Japanese giving orders. When Second Battalion was under siege at Nhpum

Ga, Roy Matsumoto crawled out as far as he could from the perimeter and listened to the commands. He reported to Lieutenant McLogan that the Japanese planned to attack at dawn and rush the overhanging projection of land that about twenty men were protecting from their foxholes. McLogan decided to surprise the Japs and pull the men back from their foxholes, which were subsequently boobie-trapped. They re-established the perimeter and dug foxholes further back, then waited quietly. When they heard the battle cry 'Banzai!' which means 'honorable suicide,' followed by cries of 'Death to Americans!', they watched as the Japanese fired into the empty foxholes and threw grenades, but the explosions that followed surprised the heck out of them thanks to McLogan's sneaky maneuver.

"Sergeant Roy Matsumoto listened carefully to the dialect of the Japanese officers rattling off their commands and realized that he knew that dialect. When Matsumoto noticed the confusion of the Japs when they realized that the foxholes were empty, he stood up and yelled the command 'Susumu!', which means to charge in their Japanese dialect. The Japanese charged up the hill, and when they were approximately fifteen yards away, the Marauders, with their tommy guns, Browning automatic weapons, and grenades, laid a barrage of lead on them. Fifty-four bodies were counted a short time later.[33] Although Matsumoto was the hero of the day and his actions became legendary, the Marauders of the Second Battalion had to fight for six more days."

April 7, [1944]. Up at daybreak. Khaki is pushing on thru us. Orange is to protect trail. 4th platoon is to follow Khaki and secure trail. Please God guide and protect us in our coming trials. Thank you for keeping us safe up to now God. The 1st sergeant was killed. The other sergeant wounded last night. 4th [Platoon] was commented on. Fab job we have done and 1st squad as point. Dive bombers, Artillery and everything hit 3 times & we attacked 3 times. We finally took first part of hill. Moved back to perimeter. Had tea, bouillon, and cheese. Feel swell. Dirty and contented. Had cgts and rats issued. Expect Nips to move out. Bed early. Slept well. 1 hour of guard.

Success in gaining ground made all our efforts worthwhile. It perked us up a bit. We also heard that 1st Battalion made it to the airstrip at Hsamshingyang, which really raised our spirits.

We were determined to break the siege that held 2nd Battalion captive.

April 8, [1944]. Up and at em at 5 a.m. Colder than hell. A bat is going to right. Khaki up the trail and we are going around left flank. Started at 8 a.m. Boy, its rough going thru jungle up hill and down. Finally bivouacked on nose of hill overlooking town. B Bn is in. Have perimeter. God, please guide and protect us. Carry us safely thru this campaign and back to the States for Evelyn's sake. In Jesus name, Amen.

April 9, [1944]. Up 5:30 a.m. moved S. and west to hit Nip flank. Jungle is rough going all day long. Reached our place at 2 p.m. Found out that Khaki had moved up within 75 yards from B Bat. Moved back out again. Uphill and down and contacted Nip patrol. Moved back to trail and took up old position to protect trail. B Bn got thru and took out 147 wounded and 36 killed. Rough as the devil. Spread out thin.

"April 9, 1944, was resurrection day. It was Easter Sunday. It was resurrection day in more ways than one. The siege of the Second Battalion at Nhpum Ga that had begun on the night of March 28, 1944, was lifted, and the men of the Second Battalion were rescued. Everything that had looked so dark and almost hopeless at that point with the Second Battalion's inability to move and establish a strong defensive position had become a bitter battle in which 57 men were killed, and 302 were wounded in action, according to the Military Intelligence Division of the U.S. War Department, which included the unit's movement toward Inkangatawng to the end of the siege.[34] The enemy losses totaled 700. Ninety of those enemy losses were eliminated by the action of Weston's and our platoon, under Lieutenant Warren Smith, during our blocking and withdrawal action. The scouts from the 3rd Battalion found the 2nd Battalion's perimeter through a horrific, bombed, and strafed landscape made more surreal with dead Japanese everywhere. Colonel Beach radioed Colonel McGee, who was in the perimeter of Nhpum Ga, to fire three shots. The scouts were close by, and a few minutes later, they and Major Briggs of the 3rd Battalion's Kahaki combat team entered the Nhpum Ga perimeter with its nauseating stench of death.

"Bloated animal carcasses and dead Japanese were strewn about. A cleansing breeze from a kettle of Himalayan vultures swooped down, and now that the chaos was quieted, a wake of those huge meat-eating vultures and bluebottle flies were feasting. Bigger than houseflies, the flies sucked the living life out of the surviving, dehydrated mules and horses. Those flies looked like a metallic-gray, chain mail armor covering those poor creatures. I imagined an army of fungi, which included hundreds of types of molds, mildew, mushrooms, smuts, and rusts, beginning their webbing of digestion, a parasitic relationship with the dead. I thought how amazing the earth was that these living organisms performed a burial service to break down the dead and return it to humus—that dust to dust of humanity became hallowed ground that we fought on. It reminded me also of what Lieutenant Weston said, that 'war was not about real estate, but power.'[35] It must have been a searing sight of the Marauder survivors who were emaciated and dehydrated yet assembled and walked out with their dead and wounded. They were the dead carrying the dead. Their faces were expressionless, and their movements robotic. Their sunken, ashen faces and hollow, bloodshot eyes stared in disbelief that they had been called from the tomb."

April 10, [1944]. Up early. Sergeant Mason has strong point. Went out on patrol S. & West. Went out 2 hours. Saw jap tracks at water point. Rough as the devil. Found Lt. Smaules bivouac area for last night & his tracks heading south. Went up to Nhpum Ga and got extra rat for men & saw dead horses and Japs all over. Area is shelled all over. Mulligan and Smith think they are dictators. Not worth a shit. Tonight, we are on A strong point. Just 6 men 250 yards on each side. Rough. God, please protect and guide us safely thru this. It's going to be rough going for Evelyn's sake in Jesus' name. Amen

April 11, [1944]. Up 5 a.m. Eyes and ears open. This outpost is dangerous. Ate good breakfast. Egg yolk, crackers, and coffee. I pray to God to be relieved before the day is over and please God protect and guide us safely thru this. In Jesus name, Amen. Platoon of Nips approaching us from the East. They were bombed and strafed. B BN pulled out and B.C.T [Blue Combat Team] and ABN was put in. Everything looks black. 2 hours on outpost 12-2. Ate cheese and lemon and lots of chow. Need bath. 2 weeks since I had one. Lots of noise. Think Nips occupied positions 75 ft across trail. Not any sleep.

April 12, [1944]. Up 5 a.m. Rolled packs. No Nips, I guess. Got radio fr BN called our patrol every 2 hours. Washed face, teeth, hands. Ate cheese and coffee. Outpost with Smuck. 10-12. Mason was busted. Good boy screwed by Mulligan and Smith. No good people. Pulled in outpost 4 p.m. Everything okay. Go on patrol up trail. We are going to be relieved on 15th. It's wonderful or at least, I hope. Ate bouillon, cheese, and coffee. Bed at 7[:]30 [p.m.] Guard 8 -10. 3-4 gets kinda rough. No sleep. Mahmood went to B Battalion hill. Got medicine for leg sores. Sent up 1 day's ration. No water as of yet.

April 13, [1944]. My bad luck day. But I'm praying to God to see me safely thru this. Goddard and McBride went up to hill. It looks like rain. Rained last nite. Kinda rough. Taking it easy. Goddard and I went to Chang Tang yang [Hsamshingyang] Air strip. Got 3 letters fr Evelyn 17-28 Feb. and Mar. 10. Pepped me up. Wonderful. Got 4 carriers, canteen for squad. 6 toothbrushes, shoes, and everything. Tired as hell climbing hills. Saw Peters, Pongrata, Chambles, Maddry from Battalion. 7-3 guard duty. Tired as hell.

April 14, [1944]. Hit the deck very early. Mahmood and I went down and got water then made patrol up to Sgt Oliver's. Gorretti came by and tried to fix radio.

Mahmood went down to air strip to get supplies for us. Rumored we are to get Reld today. This outpost is hell. Didn't get relieved. Got ground coffee, tea, sugar, etc. Recd 4 letters fr Evelyn. 1 fr Kat and 1 fr Howers. Was grand. Couldn't be better. God is kind. Good rat cheese, eggs, coffee. Went to B BN hill. Will be Reld tomorrow. Swell. Tired & sore & dirty. Bed 8 p.m. Guard 10-12 p.m. God please guide and protect us thru this for Evelyn's sake in Jesus' name Amen.

April 15, [1944]. Hit the deck 5 a.m. Had trouble w/ McBride. He turned chicken and wouldn't fight. Ate same old stuff. Cheese, Bouillon, Crackers. Guns blasted loose. Expect to be relieved today. God, I hope so. This is getting on my nerves. Got Relieved by Red C. T. [Combat Team] at 1050. Moved up to maggot hill was put in perimeter by Sergeant Mulligan. No good. Made bunk w/ Goddard. Got water. Met good friends from A BN C. Co. Ate chow. Saw Doeling. Bed early. ART opened up. Lots of hand grenades.

"After seven days, we were relieved from that hell of a dangerous outpost that Sergeant Mulligan put us on. That perimeter consisted of over a quarter of a mile, or 450 yards of dense jungle, not open fields. We were only one squad of six soldiers. The nips were 75 yards away. None of us could sleep at night, even though some of us had guard duty, eight-hour guard duty at that. Fear is a human reaction when we feel out of control, especially when we can't see. Not having our sight is very frightful, like a jungle at night, but if you can plan an escape route in your mind, it calms your survival-screaming brain down a little. After being in combat, you get used to the noises and learn to think under those circumstances, but you never get complacent or forgetful of the miracles that take place when you survive a close call."

"Dad," I asked, "it didn't make you paranoid that Sergeant Mulligan was trying to get you killed? Didn't they know that an assignment like that for that length of time diminishes morale?"

"The men were grumbling as well, but I don't see how that would have benefited the situation if any of us had been killed 'cause the Japs outnumbered us. I think he felt safer because our squad was on guard duty. It's like riding a good horse all the time; pretty soon, it will wear out. However, it was terrible for morale and the psychological implications—like when McBride turned chicken. He was shaking and couldn't stop, as many of our soldiers did at one point or another. I had to really talk to him about how important he was to all of us surviving. We all had to be a team and fight our way out. We were relieved but sent to Maggot Hill, where we were surrounded by dead Japs, mules, and horses even though lime was put on them. This was purposeful by Sergeant Mulligan. Later, I reported this incident to Captain Haol."

> *April 16, [1944]. Up 830 a.m. Nice sleep. Quite a bit of excitement last night. Fired 4 rounds. Lots of rumors. Saw Hickman, Peters, and the rest of C Co. men. They certainly like me. Cut the squad's hair and washed up. Got a little rest. Think we have the Nips on the run. Thank you, God for guiding and protecting us. Keep them safe at home. Dug communal trenches. Bed 7 p.m. on guard 2nd. Still lots of firing.*
>
> *April 17, [1944]. Up 7 [a.m]. Still nice sleep. Went on patrol out in front and picked up Jap helmet,*

cartridge belt, bayonet and scabbard, canteen, and mess kit. Cleaned them up and disarmed hand grenade and knee mortar. Washed up. Got 10-1 rat. Saw Mason caught hell from that 8 ball Mulligan. Wrote V mail to Evelyn. Swell to hear from her. Bed early. Tired.

Oscar, Mahmood, Gorretti, McBride, and I sat around eating our K rations. There was a dead Jap face down with one leg bent at the knee. He was in a state of *rigor mortis*, so the bottom of his tabi, which was a rubber boot with a split toe, made a little side table. Oscar rested his canteen on it and called it his 'Japanese foot stool.' We got a laugh out of that. We proceeded to talk about our favorite subject——food, specifically K rations, as if they were a culinary delight.

Oscar said, "I'm at Patsy's Restaurant in New York City right now, and I have been served a rack of lamb, Italian style, with lots of garlic. Frank Sinatra, old Blue Eyes himself, sat down at the bar and ordered a double Jack Daniel's."

We laughed.

"My steak is cooked just right, a little rare on the inside. Mm, and my beautiful bride just served it to me. She has homemade vanilla ice cream with chunks of pineapple for dessert. How is yours?" I asked.

"I'm at our local diner, and I just put a nickel in the old jukebox, and 'Boogie Woogie Bugle Boy' is playing. The waitress just served me barbeque chicken with a side of baked beans and potato salad," McBride said.

"You guys don't know what's good. I'm eating veal

parmigiana just like my mother used to make," Gorretti said.

"Tony butchered a mule steak that just came off the grill, and I put butter on top of mine. You didn't get any?" Mahmood asked.

We grinned and laughed about our meal; it put us in a better mood because we had missed so many days of food. We were grateful for anything we could get. The pilots who flew the Hump over the Himalayas to get supplies to us were courageous and undeniably a major contributor to our victory in Burma. When there was high wind and rain, they couldn't fly, and we didn't eat.

"That's a cute story," I said, "but I am amazed at how desensitized you guys became to the point that you could eat around a dead human being. I don't think I would have an appetite. When I toured the Nazi concentration camps of Dachau, Auschwitz, and Majdanek, I learned what the victims of concentration camps went through. They could sleep next to a dead person with lice and feces and think nothing of it."

"It's incredible how a human being can get used to just about anything when your survival is at stake," Dad agreed. "I know because I lived it."

"The difference was that the concentration camp inmates had no power, were not united, and were demoralized to the point that they turned inward and became unaware of what fellow inmates were going through. I've been reading a book by Dr. Victor Frankl entitled *Man's Search for Meaning*. He was a Jewish psychiatrist who was imprisoned at Auschwitz," I said.

I leaned over and pulled the book from my tote bag.

"Let me read something that's interesting. World War II wasn't just about fighting the Japanese, it was about fighting the Nazis that had expanded all over Europe as well. Dr. Frankle wrote, 'I refuse to call people collectively guilty. There is no collective guilt, it does not exist, and I say this not only today, but I've said so from day one when I was liberated from my last concentration camp. Guilt can, in any case, only be personal guilt—the guilt for something I myself have done—or may have failed to do! But I cannot be guilty of something that other people have done, even if it is my parents or grandparents. And to try to persuade today's Austrians between the ages of naught and fifty of a sort of 'retroactive collective guilt,' I consider to be a crime and an insanity—or, to put it in a psychiatrist's terms, it would be a crime, were it not a case of insanity. And a return to the so-called 'kin liability' of the Nazis! And I think that the victims of former collective persecution should be the first to agree with me. Otherwise, it would be as if they set great store by driving young people into the arms of the old Nazis or the neo-Nazis!'"[36]

"So," Dad said, "in the case of the Marauders, we volunteered for the job, although many times we felt trapped. In the last dying days of the Marauders, we were fighting for each other to survive. We had weaponry that gave us the power and courage to fight against injustice, to fight against a deadly enemy, to fight against an 'insane enemy' as Dr. Frankle mentioned. God will judge us as individuals."

"Exactly," I agreed. "In simple terms, judging by generalization is unjust. It's like the old cowboy movies that depicted the good cowboy as the one with a white hat and the bad one with a black hat. That's ludicrous - if not laughable! God will judge each individual person by their own personal intentions, but the law would probably put more emphasis on the outcome."

April 17, 1944

Darling Evelyn,

I now have a chance to write, so will put what little I can in this V-Mail. I am somewhere in Burma fighting alongside of the Chinese forces. Am feeling swell, as good as could be expected and sure hope everyone there is okay. I can say one thing, your 64 letters certainly raised the old spirit. I can't explain how much they helped me, but just saying thanks is a far cry for how I feel.

I am sitting on Maggott Hill with the forever famous Merrill's Marauders. I miss and love you. Oh, how I wish I could have stayed with you. When this is over, I am coming back, and we can carry on from where we left off. I love you and adore you with everything I have. You are my life, my everything.

Love,
Larry

April 18, [1944]. Hit the deck 8 a.m. Good breakfast. Went to 3rd platoon. Am going to see Captain Haol. Watch fixed by Blue. Were issued mountain rations. Swell. Washed and shaved. Wrote letter to Evelyn. Am full. Hope to go to company headquarters. Bed Early. Saw Captain.

April 19, [1944]. Hit the deck. Ate dried beans and coffee. Not bad cereal. Smuck's squad went on patrol with red [Red Combat Team]to find Hick. Orange column is going to air strip is swell. Left 11 a.m. Arrived 2 p.m. It is wonderful to get the rest. Pongratz outfitted me in clothes. Swell. Saw Maddry. Sleep early after bath. Mahmood transferred to 3rd platoon.

"You had dried beans with coffee like cereal?" I asked incredulously.

"Yeah, it was crunchy!" he said and paused. "I don't know why Mahmood was sent to 3rd Platoon. Maybe because of the losses that the 3rd platoon experienced. Mahmood and I had each other's backs. Good man. Good fighter. I missed having him around. Still checked on him, but it wasn't the same."

April 20, [1944]. Up Early. All of squad got new clothes. Was issued 10-1 rats Would love to see my Honey. God thank you for carrying us safely thru. Please God continue. Rumors everywhere. Don't know what we are going to do. Nice chow. Was also issued J rats. Nice Took bath. Made Lt. Smaules some candy. Bed early.

"I received mail from Mama."

Dear Son,

My precious boy, you have been on my mind all the time. I am on my knees for you as I was for your brothers. You know Oscar was in Guadalcanal. He was in his foxhole when he was hit in the head by a large piece of a Palm tree when a grenade exploded near him. It knocked him out cold, so when the Japs came around to kill anyone that was still alive, they skipped him thinking he was already dead. He's okay but quiet and brooding. I don't know where you are in the world, but you are in my heart.

Mamie had a vision of you. You and your fellow soldiers were in the jungle when the Japanese soldiers were moving in. Little did you know that you left at the most crucial moment, and you left in a hurry when you realized how close the Japanese were to you.

Mamie also saw you carrying a rifle and running across a river. The enemy was firing at you. You made it across the river and ran along the banks until you passed out. A soldier threw you across a horse and tied you to it. He jumped on the horse with you in front of him. He kicked the horse, and it took off in a wild gallop. He got you to safety.

I talked with Evelyn on the phone. I'm sure it cost her a pretty penny, but she said that she didn't care. I told her about Mamie's vision. We have contacted all of our pastors, churches, friends, and families. We are calling the Lord's Army to rescue you and to defeat the Japanese so that this war will be over. Son, pray out loud, so the angels can hear you! Remember Mind over matter!

Love,
Mama

"I pulled out a piece of paper. I was stunned. All I could write was what I wanted her to know."

Mama,

Tell Mamie that she was exactly right. She must have been there with me. We need an army greater than us.

Love,
Larry

"Wait just a minute," I interjected. "Would you repeat that?"

Daddy read the letter again.

"Aunt Mamie dreamed this?"

"No, she did not dream it. She had a vision of the whole scene. She had been praying for my safety and saw the Japs that outnumbered us creep toward us like in a surprise attack. We got the order to move out fast, and she saw me running through the jungle and across a river while being fired on. She saw me faint after running for miles and being thrown onto a horse. After the war, I talked with her about it. She said that visions happen when she was tired and not focused. It seemed to happen when one sense didn't overpower another one—when all of her senses were equal."

"That is incredible," I said. "So, what can she do about it when it happens?"

"Nothing but pray."

> *April 21, [1944]. Friday. Up 5 a.m. Reveille 6 a.m. 2 miles from Japs and behind their lines. Phooey. One hour of close order drill and then BAR then full field inspection, but it rained us out. Got wet as hell. Saw Pongratz. Got rations C. Chinks come in tomorrow. We are moving out tomorrow. 1 ____ Bed Early*

"We were exhausted, and the officers wanted us to practice a close-order drill in the jungle. That beats all of the numbskull, asinine orders I had ever heard of. The average weight loss for the men was thirty-five pounds, but I lost fifty. We needed rest. We needed good food to carry on with the fighting. We did not need to expend our energy on a drill. I realize that the drill was a bonding, morale, and teamwork exercise. We had already bonded with our brothers-in-arms. Morale was the lame horse they beat to make it to the finish line. We started our drill, but fortunately for us, it rained, and it was called off. I was glad for the rain this time," Dad said.

> *April 22, [1944] Saturday. Hit the deck. Have very bad chest cold. Lots of chow. Got some Sulpha diazine [sulfadiazine] tablets. Will move at 1 p.m. 800 yards. Chinks to take over. Am sick and tired. God, please help us, guide us safely and protect us thru the coming storm. [_?_]moved to new area. No perimeter. Nice area. Mac and I got water. Saw old pals from other plt. Bed at 8:30 [p.m].*

> *April 23, [1944]. Sunday. Up 6 a.m. 1-man water guard. Good chow. Bacon, roast beef. Chinks moved in area. Goddard was sick as hell last night. Got cigs*

traded for chink, Jap, Indian and Ceylon Money, Clark and I got [_?_] for platoon. Rested good. Moving out tomorrow.

April 24, [1944] Monday. Up early. Moved out at 8 a.m. All uphill 15 miles in 8 hours. Rough. Into Wellington at 5 p.m. Wet w/ sweat and tired. Ate chow. Had water guard. Made Staff Sergeant 16th of April. 300 Japs 1 days march S.E. Tied in with 88th Chinese Regiment. Good looking boys. Bed early. Tired & sore.

"I was on water guard around dawn, first light, and standing very still in the shadows of the trees. I was alert to any movement that may be that of the Japanese, but I did enjoy the dappled light coming from the jungle sky. I looked forward to the day I could look up at blue skies or see the stars. I was startled by the smell of popcorn, but it was from the urine of a magnificent tiger. It came out of the brush and turned its huge head with the three black horizontal lines and one black vertical line down the middle of its forehead and stared at me. I stared at him. I didn't move, not even my eyes or eyelids. He was across the small water hole, and I knew with one leap he could kill me, but I could shoot him as well. It was one of those moments when I was awestruck by the beauty of this creature. I knew in my heart that I could not kill him. I wondered if I would ever be able to look at a Japanese and think the same of him. The tiger stared at me warily, then crouched down close to the water and drank its fill. I told the guys when I got off duty. McBride said, "Maybe you smelled so much like the jungle that the tiger didn't notice." The other guys laughed and made jokes about popcorn and the tigress having popcorn babies. Those crazy fools even put mud lines on their foreheads. I told them that in Chinese, that was the pictograph for the word tiger. It also meant

King, as in "King of the Jungle." 'Yeah, we are tigers, alright, we are going to claw the Japs to death.' Kings and Tigers, I thought. Fierce, yet fragile rulers and warriors like Galahad."

I smiled at Dad's comical memory of those crazy guys. I also remembered an event when I was a child of about eight years old that was not comical. Daddy was moody, quiet, sullen. I thought he was mad at me, but I didn't know what I had done wrong. I walked into the kitchen to the utility closet door. It had slats in the door for ventilation that the orange, setting sun poured through. The door was ajar, and Daddy was eating a slice of watermelon over the trash can with a butcher knife. The orange strips of light were that of a tiger. The shadows looked like black stripes. The red melon at his mouth looked like blood. He was vicious-looking.

The door squeaked when I opened it a little wider. It must have scared him because he jumped and raised the knife; upon seeing me, he deflected the knife and jammed it into the door frame. He was incredibly fast. I stood my ground; my blue eyes shot tracers at him. Instinctively, I backed out and turned from him as if he were a tiger in the wild. I knew I could not run, scream, or make a sound. I walked slowly across the black and white squares of the kitchen floor and out of the door.

My heart was pounding when I made my way outside. I sat on the front porch steps, holding my knees and rocking back and forth. I wouldn't let myself cry because Daddy had told me one time that I had to practice bravery even if I didn't feel it at the time. 'Practice, and you will be brave,' he said. I never knew, months before this incident, that I would have to be brave around my own father. I knew my father loved me; he did everything for me and my siblings. He built playhouses, a basha hut, tree houses, and rope swings. He even built us a go-cart, but why did he threaten me? I contemplated telling my mom, but I knew it would cause trouble, so I didn't. As an adult, I realized that I had done nothing wrong as a child. Dad was in a flashback horror of the war, and I had interrupted it. He reacted as if his life was threatened, and I had been a Japanese soldier.

That memory birthed another memory of a frightening event between us. I was a teenager, probably around seventeen. I had fussed with my hair and makeup and was proud of my new dress. I had planned to meet my friends for a fun Friday night when Daddy stopped me in the foyer.

"You are not going out in that dress. It is way too short," he said.

"But Daddy, this is the style. It's a mini skirt. All the girls wear them. It's the latest rage. Haven't you seen them on TV?

"Do you think that because it's on TV it is okay? You don't take your values from TV, young lady. Now go change if you want to go out."

"No," I said.

I clenched my jaw. I was as stubborn as he was. Daddy was standing in the hallway to the foyer–about 4 feet from me. When I defied him by saying 'no,' I was reminded of the tiger in the utility closet that I witnessed as a child. Dad made the meanest look, and he could make mean looks that scorched you.

"Now!" he said with the intensity of a drill sergeant.

I thought he was going to escort me to my room. I didn't move and came to regret it. I saw flashes of orange and black stripes—red blood at his mouth. I backed up and bumped the small foyer table and blindly reached for a heavy crystal ashtray that my aunt Norma had given to my mother. I tried to protect myself, but Dad's military training kicked in automatically. In a flash as fast as gunfire, Dad was there with one hand at my throat and the other disabling my throwing arm. The ashtray fell to the floor with a loud crash, and I screamed a glass-breaking soprano louder than I knew my voice could make. I had forgotten all that I learned about the tiger as a young child. I thought he was gone all this time. I confronted him and was taken down in a heartbeat.

Then my mother was there, looking bewildered at the screams and loud, crashing chaos she heard but had not witnessed. She admonished me for breaking her crystal ashtray. I ran to my room and flung myself on the bed. Several minutes later, Dad came in and knelt by my bed. He checked my little finger that was jammed in the fray. He was saying words to me; I think it was some sort of apology, but I couldn't respond. I was too angry and frightened. I nodded so he would go away. I didn't get to go out that night, and I was mad.

The next morning, Dad was eating breakfast at the counter, and I was sitting at the opposite end, ignoring him. I was still angry and didn't look at him. His cereal crunching irritated me. I tried to enjoy the aroma and taste of my coffee, but to no avail. I felt like a porcupine with quills bristling from every pore of my body. Daddy's chair repeatedly squeaked, and I looked over at him. He was shaking from suppressed laughter and then laughed out loud a hearty laugh.

"You are a mean little cuss," he said between bursts of hilarious laughter. I started laughing, too.

"Like father, like daughter," I said. The quills flattened the more I laughed.

Years later, I realized that Dad's military skills must have been incredible because he was agile and extremely fast. He could control his emotions, which was a lesson to me. I also realized that Dad's survival was a cooperation with God.

I remembered all of this in seconds, and it played out so real in my mind.

Now I understood Dad's memories that stuck like tar in the brain. Impossible to forget. I lamented the fact that there was no support or psychological help for him back then. I was like him in some respects. I felt that I would get over that incident with a little time, and I did because I was strong-willed like him. There was no *kumbaya*; smoke a little pot and get over it. No, Dad and I would never use drugs or alcohol to medicate our feelings. If we were going to get over it, we would do so with the clarity of a thinking brain.

April 25, [1944]. Teus. [Tues] Up 5:30 a.m. Chowed up. Slept well. Rolled pack. Are moving 7 miles. Uphill is rough. Leave out at 9 a.m. Rain and sweat. 4 miles up, 3 down. Chinese in front A BN. Taylor is a damn sponger. Wet and tired. Bivouacked. Ate good. Bed early. Slept well.

"Dad, did you not like Taylor?" I asked.

"Oh, of course, I liked Taylor. We were brothers, and just like brothers, you sometimes get into squabbles. We had each other's backs. I can't even remember what it was about. I was close to everyone in my squad. Those in my squad, which was 1st squad, were: T/5 Hayward McBride, Pfc Albert Mahmood, Pfc Harvey Taylor, Cpl John Smuck, Pfc James A. Gorretti. Other significant friends from other squads were Goddard and Mattlock. Mattlock was a good friend."

> *April 26, [1944]. Wednesday. Up early ate K rations and ready to move at 8 [a.m.] Going from 1 to 4 miles. Very hilly to Nabum. Divisional HQS. The 3 miles and arrived at 10 a.m. 1st Sqd was point. Bivouacked. Took bath. Went on patrol. Found Nip trail. Was rough. Mac got fish. We cooked in bamboo. Bed 8 [p.m.] Tired and sweaty.*

"Our outfit was one of surprise attacks, and silence was ordered, but as often as we could, when the noise was not a factor, we would throw compression grenades into the rivers, which usually yielded about twenty fish. We also shot game when there was no food."

"Isn't there a fuse on the grenade that would get wet?"

"No, the ignition system is internal."

Myitkyina

April 27, [1944]. Thursday. Was Trfd to 1st platoon. Whole squad. Swell to leave Mulligan and Smith. Moved out at 8:30 a.m. Up a rough hill. Hot as hell. 4 miles into bivouac. Took bath, washed clothes and read Omnibook. Smoked pipe. Drew 2 days rats. Mac and Snyder made candy. Bed early. Made banana shelter.

"The number of soldiers was so diminished by those who were killed or wounded as well as those who had diseases that Headquarters reorganized and combined their resources. Each battalion had 600 men. They were H Force under Colonel Hunter, M Force under Colonel McGee, which included three hundred Kachins, and K Force commanded by Colonel Kinnison. The Kachins were more valuable than we expected. They were always there to guide us, but they were more valuable in patrolling trails, making ambushes, and booby traps. Most of all, they were superb guerilla fighters and gave the Japanese the impression that our outfit was much bigger than it was. My squad and I were moved to the 1st platoon, which was swell. I think it was the result of my meeting with Captain Haol. I was now in K Force, which combined the Chinese 88th Infantry Regiment 30th Division and the Marauders of the 3rd Battalion.

"We had heard rumors that we were going to have a 3rd mission to take Myitkyina and, yes, the rumors turned out to be true when we were briefed with very little information except for the promise of flying us out at the Myitkyina Airstrip and flying in more Chinese troops and supplies. We took that promise with a grain of salt. All the guys were more than disgruntled when they heard the news. General Stillwell never visited us, as we expected of a General after what we had been through to take Nhpum Ga. Some of the men were so mad they theorized that maybe Stillwell was afraid that he would be popped off if he visited us. Who knows?

"We were grateful for the Docs taking up for us and telling the upper echelon that we were broken, underweight, malnourished, dehydrated, and with various diseases like malaria, amoebic dysentery, skin rashes, infected sores from leeches, and the deadly scrub typhus. Stillwell had an ego to bolster. He didn't care. We had fought at Nhpum Ga at 2,400 feet in elevation, and now they wanted us to march over the Kumon Mountain Range, which was about 6,100 feet; however, we didn't know that until we were doing it. We were heading to Myitkyina by way of Ritpong, which lay on the other side of the highest Kumon Mountain range through the Naura Hkyat Pass. It was more than the marching; it was marching in inclement weather. The monsoon season had begun. It was common for the rain to last ten hours; however, the intense, hard-hitting rain lasted about two hours. I later learned that this mountain was part of the Himalayas and was known to be infested with disease, especially the larva mite that transmitted typhus.

April 28. [1944]. Friday. Up 5 a.m. Rolled packs. Moved out to Nabum then E. 9 miles in all. Saw all old boys from A BN. Hot as hell. Went up hill. 7 miles. Getting rougher. Wore out. Moved into bivouac area at 530 [p.m.] Had 1 drink believe it or not. Made candy. Rained. Bug bite all night.

April 29, [1944]. Saturday. Up cold as hell. Clark and I got water out of mud hole. Ate breakfast. Moved out at 8 a.m. Marched up hill 3 ½ miles. Boy its rough. Rained during dinner break. Miserable. Lost 12 animals from landslides. Bivouacked at edge of trail. Chinks behind. Khaki in front. Pack is lighter. Slept with Goddard.

April 30, [1944]. Sunday. Up and at em. It rained 7-830 [a.m.] Uphill mud and slippery. Had to dig out trail. Moved 4 miles. 1st plt is in the lead. Just one more big hill after this one. Made dextro candy good. 14 horses went over side of hill. Bed 730 [p.m.] Rained all night.

"I had a soft spot for the animals. I hated to see fourteen horses slide off the mountain. Of course, we had to retrieve their packs, but in some cases, we had taken their packs off to enable them to climb the slope easier. It was a shame that they had made it through a hail of lead from guns to shrapnel, only to fall off the side of a mountain. It happened so fast. Everything in war seems to happen at lightning speed—it takes your brain a few beats to gather your wits about you. I checked with the muleskinners to see if Red was okay when I had a break. She was okay, and I said a prayer for her to be sure-footed through the rest of the campaign. The exhausted Marauders had to carry the contents of the packs the animals carried."

May 1, [1944]. Monday. Up 530 [a.m]. Ate breakfast. Rolled pack. Moving out 730 [a.m.] Reached top of hill 10 a.m. Boy it was wonderful to walk down again. Muddy as hell. Rained all day. Dug new trails all day. Rough Reached bottom 4 p.m. Bivouacked. Nice supper. Bed early Muddy & tired.

The sun was going down, creating ethereal light through the branches of the oak trees around Mom and Dad's house. Dad and

I sat outside behind the house and enjoyed the sun preparing for slumber.

"I think I understand when you talked about rhythm and being in the moment," I said. "I remember when I rode Little Joe out in that backfield. I galloped him and got into his rhythm— the rhythm of the gallop. When I was perfectly balanced and moving with him, it felt like I was an extension of him, or he was an extension of me, so much so that I could lift my head to the sky and stretch my arms out to the side. It was glorious, almost divine. However, I never did that when I left the barn because several yards out, Joe would do a U-turn and head right back."

Dad chuckled, "I remember that."

"I knew about this flow when I was a kid. I was completely immersed in the moment and completely focused. I couldn't explain it, but I understood it. I also experienced synchronicity. I didn't know the term for it back then, but I knew about coincidences. I remember sticking my hand under the bed, and a splinter from the bed slat went under my thumbnail for its entire length. Then I saw you that weekend, and you also had a splinter under your thumbnail, just like mine. I think you were working with lumber when you got yours. Go figure. There must be a message somewhere."

"We should stay away from lumber," he quipped.

"You are a pain in my thumb," I laughed.

"Ditto," Dad said.

Seriously, now, as an adult, I've come to understand that real knowing is when you are still, quiet, and meditative. You know

God is there. He's almost tangible. Maybe you experienced that flow when your unit was fighting together."

"Yeah, like a well-oiled machine and God was operating that machinery."

"Dad," I said quite unexpectedly. "I have a confession to make. As a child, I climbed into the attic and found your chest of war stuff. I read briefly in your diary that you ate a rat. That evening at the dinner table, I said that the spinach tasted like a rat, and you asked, 'How do you know what a rat tastes like?' Then I asked, 'Do you?' and you said 'no'. I was puzzled about what you wrote in your diary."

Dad started laughing, "Rat was short for rations."

"I know that now, but I didn't know that then."

"You weren't ready to hear about all of that back then."

The sun was sleepier than ever and seemed to seek the cover of the earth. It was the perfect time to rest and enjoy the beauty around us, although my shoulders sagged at the effort and weight of the war that Dad shared and I recorded. I was enthralled with my dad's diary and memories, but I also wanted to be considerate of him.

"Dad, do you think we have covered enough for one day?"

"Yes indeed, let's take this up tomorrow. I will see you in the morning if I can get through the jungle of nightmares," he said.

The next day, my father brought coffee upstairs to my mother and me. This was a sweet gesture that he did for my mother every day when I was growing up and every day of their marriage. The three of us sat out on the balcony sipping our coffee and talking about their garden, the pear and fig trees, and the work it takes to make fig preserves. Then, we sat silent and opened our minds to the possibilities of the new day.

"Did you sleep well?" I asked.

"Other than dodging bullets, I slept like a baby," he said.

"If you two are going to get that diary finished, you need to get to marching," Mother said with a laugh.

"Yes, drill sergeant," Dad and I said.

As she carried the tray downstairs, Dad and I saw a black sedan turn down the road to Rich and Satchiko's house.

"That's odd," Dad said, "we don't see that type of car in this area very often."

I gathered my recorder and went downstairs to begin another interview session. Dad read from where we left off.

May 2, [1944]. Teus [Tues] up 5 a.m. Ate frugal meal. Mac went to sleep on guard. It's bad. Have 6 miles of bad trail today. Muddy. 7 a.m. marched 10 miles up and down. We were rear guard. Arrived 5:00 p.m. Slept on side of mountain. Kachins contacted nips.

"I later learned that the trail had been out of use for ten years, and the natives never used it during the monsoon season. I'm glad I didn't know that then."

May 3, [1944]. Wed. Up. Sick call at 7 a.m. May stay here today. Have air drop. Slept well. Leaches are bad.

Went to within 300 yards of nips. Kachin scouts came back & put up outpost 1500 yds down river. No food for 2 days. Ate fish and rice. 2-man guard tonight. Bed early.

May 4, [1944]. Thurs up early. Moved to top of Hill. Drew 4 days rations moved out at 7 a.m. Followed Chinese. Marched 10 miles passed Khaki. This is a big drive Chinese, British Ghurkas, Indians, Kachins to Mishinau [Myitkyina]. Rough. Made candy. Bed 8 a.m.

May 5, [1944]. Up at 5:30 [a.m.] Raining left out at 8 a.m. 4th squad is point. Expect Japs. can't tell. Marched 6 miles muddy, tired & sore. Ate some rice. Not bad. Ate some rice for dinner. Waited for I & R report on nips bivouacked at hill. Bed 830 [p.m] after talking to Kasuly.

May 6, [1944]. Sat. Up Early. Strout went to sleep on guard. Bad. Men are wore out. Leaving out early moved out at 7 a.m. Break and march. Contacted Nips. Flanking them. Moved 6 miles. Bed after dark. Rained.

"The guys were exhausted. Falling asleep on guard duty could mean the death of all of us and jeopardize the whole mission." Dad said. "It made it harder for me to sleep at night when I was worried that someone would fall asleep on guard duty."

May 7, [1944]. Sund up 5 a.m. ate & Taylor moved w/ radio. Rain & wet. Marched 5 miles. Rained. Squad stole extra rations and tea. Swell. Nips are firing to beat hell & we bivouacked. Bed early. Guns all night.

May 8, [1944]. Mond. Up 5 a.m. cleaned guns. Firing all around us, ½ hour guard. Ate good meals. Air drop [_?_] layed around. 1st BN is pulling ahead. Good. Another Chinese Regt pulling in. God, please guide us safely through this [_?_] Amen.

"I remember the last time I saw Red. I rested for a while and walked over to visit her and J.D."

I rubbed her down with some hay and gave her a sugar cube. I whispered in her ear.

"You're going to make it, girl," I said. "You're young, and you're going to have some foals and live in a pasture with a lot of green grass to eat."

J.D. sat near where she was tethered. He appeared to have no energy, his body slumped, and he stared out of complete exhaustion.

"How you doing, Buddy?" I asked.

"Barely making it," J.D. replied.

I sat down next to him.

"Did you hear about the mule in Second Battalion that fell off one of those high ridges?" he asked.

"No. Lots of animals go over," I answered.

"This one fell about seven hundred feet and flipped about three times in the air. He ended up falling into some bamboo on his back with his pack still on. The bamboo must have cushioned his fall because when the muleskinners got down there, they got him out, and he was ok! Can you believe that?"

"Wow, that mule deserves to survive."

"He was carrying medical supplies, so a lot of men survived too, I reckon, J.D. commented."

J. D. and I sat still for a while. Both of us stared into the green jungle and embraced that ripe moment that we knew would deliver soon. We knew the end was upon us, but we didn't know when or if we were going to live or die. The outcome was yet to be born. Death was a shadow away. I had to discipline myself when I thought how great it would be to be born out of the fear of dying into the life of heaven. I wasn't ready for that yet. Evelyn wanted me to live. I was going to live.

I moved my hands where the sun broke through the leaves. I made flapping bird wings like I did when I was a kid.

"You see that, J.D.? As long as that shadow is still moving, I know I'm alive."

J.D. didn't move, and he wasn't amused.

"What's going to happen to Red and the mules when this is over?" I asked.

"Not sure. There may not be any left at the rate we are going. I think the Army will send them and us back to India," J.D. said.

"Then what?"

"The vet said that they would be evaluated. Those that could be saved would be saved, and those that had diseases would be put down."

"Do you think Red will make it?"

"So far so good. She's getting as worn out as we are."

"Do everything you can to save her, okay?"

"You bet. If something happens to her, I will end up being a rifleman. I really don't want that to happen."

I grabbed a few rupees from my backpack and handed them to him.

"Here, buy her some oats when you get there."

"If we get there, I will. Why are you doing this?"

"I don't know. I guess to have something to worry about besides myself. Besides, we volunteered for this; she didn't."

> *May 9, 1944. Teus [Tues] Up early. Miss my honey very much. May move out. Sun came out. Goddard went on detail. Chinks attacked Japs. No outcome yet. Japs lost 17. Chinks 26. We moved on down main trail. Hot as hell. Bivouacked 430 [p.m.] Good sleep. Still fighting behind us.*

Dad read a few more entries, but he had to stop when the tape ended.

"I have to go into town to buy another cassette tape. I'll be back shortly," I said as I gathered my purse and took off. Mama related the following incident to me when I returned.

"Not long after you left, Rich and Sachiko's truck peeled around the corner and came up the driveway. Rich got out, went to the passenger side, and helped Sachiko from the truck. She held a tissue to her face. We got up and went to the door, wondering about the commotion. They came in crying. Sachiko could not speak,"

Mama said.

"'Army notification officer,' Rich said, barely getting it out. 'It's Tommy. He – he died of a heart attack. They found him slumped on his desk, he said and broke down in tears. We were shocked. Your dad and I gathered them into our arms and cried with them," Mama said.

"Oh," I gasped when she told me the news. "How devastating."

"Sachiko said, 'We shouldn't have to bury our children,'" Mama said. "It was so sad."

"I know," I said quietly with tears in my eyes as I remembered my friend.

I thought of all the Gold Star families that experience such painful news, although Rich said it was presented in a very kind and compassionate way. I would never want that kind of job, I thought. I also wondered about Dad's reluctance to befriend Sachiko, but that was not the case. He opened his arms and heart to them over the death of their son. So many military families have experienced such heartbreak. With that beautiful gesture of love, I hoped my dad was reconciled and healed from the terror of being around Japanese people, even though it had been so long ago.

Later that afternoon, Dad and I began to record his diary again.

May 10, [1944]. Wednesday Rolled packs, ate, got water. Still firing. Made candy. Have listening pos. A BN passed through us. It's about time. Chinese swiped rations. Expect to move out tomorrow.

May 11, [1944]. Thurs. Up early will move out this

morn. A & Chinese are in front on another trail. Left out 9:15. Marched 9 miles. Bivouacked at 4 p.m. Took bath. Feel better. God be with us. Made candy. Monsoon coming. Am Tired and sore.

May 12, [1944]. Friday. Up 5 a.m. ate 2 rats. Move out at 8 a.m. getting closer to Nips. Contacted them close to noon. 2 dead. Noe [William C. Noe] in 1st Platoon and 3 wounded. Bad business. Dug in. I got 1 or 2 of them. No sleep. God, please be with us. God in heaven guide us safely through the bad day tomorrow and please protect us for Evelyn's sake and in Jesus' name. Amen.

"We had contacted the Nips at Tingkrukawng and thought we were fighting a platoon. We probed their perimeter to find a weak spot when we realized that we were fighting a Japanese battalion. We had suffered eight dead and twenty casualties. The Chinese 88th Regiment, which we were attached to, suffered more severe losses. Some of the men were so exhausted that they collapsed. We had one more engagement the next day. You would think that May 13 would be my bad luck day, but it wasn't. May 13 was my miraculous day."

May 13, [1944]. Saturday. Up 5 a.m. very little sleep last night. Got another one of our men. Bad. Bullet hit my shoulder. God, please guide us safely through this and protect us for Evelyn's sake and in Jesus' name. Amen. Hit nips again. Was trapped for about 10 minutes. Rough. God, be with us in our hour of need. Nips have cut our back trail. Cut off. Bad spot.

"I was hit in the back near my left shoulder area. At first, I thought it was a bullet, but later I believe that it was shrapnel. It burned like hell."

I zoned out when Dad talked about the hit to his left shoulder. I remembered taking him to a cardiologist for heart issues about a year ago, and he told the Doctor about being hit by a bullet over his heart.

He put his fist on the center of his chest, thinking it was his heart, but more accurately, his heart was a little to his left.

"When I was fighting in the jungles of Burma during WWII, a round came in when I looked up from my fox hole. It hit me right over my heart. It knocked me over in the foxhole, and my whole chest area was badly bruised. It hurt like a son of a gun. Curious thing was that I found the slug, probably from a Japanese Arisaka, in my pocket. After that event, I felt that God was protecting me, and I relaxed a little. So, do you think that hit caused my heart damage?" he asked the Doctor.

The Doctor and I looked at each other and then at Dad in disbelief.

"Dad, this is the first time I have heard that story. You mean to tell me that you were hit in the chest by a round, and it did not penetrate, and you found the Japanese slug in your shirt pocket?" I asked.

"That's what happened. It bruised my chest badly, and it hurt like hell. It took over two weeks for the bruise to go away."

"That is phenomenal; you are one lucky man in more ways than one," the Doctor said. "But I don't think that is what caused your heart damage. Did you have Malaria or any other tropical diseases?"

"I had Malaria and Tsutsugamushi fever, which is a mite-borne typhus. In fact, there were only 149 cases of scrub typhus fever out of almost three thousand men. I also heard that 140 men died."

"Did you have a high fever with it?"

"Yes," Dad said. "106-degree temperature, and I had a blood transfusion as well. I also suffered a relapse and was given another transfusion."

"Fevers like Tsutsugamushi fever or mite typhus are transmitted by infected chiggers or larval mites. They have been known to cause heart damage when not treated in a timely manner," the Doctor explained.

"Dad," I said, "was this bullet that hit over your heart the same injury when you said your lower shoulder was hit in the May 13th diary entry?"

"No. I was hit in the heart when we were fighting at Wallawbum. My diary was to help me to remember the events if I survived. It helped keep me focused. Sometimes, the fighting was so intense that I didn't have time to write. But getting hit in the chest by a round was seared in my memory. I knew I would never forget."

"But I didn't see any entries about it in the diary at Wallawbum," I said. "I know Nhpum Ga was the most intense fighting, where you did not get any breaks. It was probably during the two days that you didn't have the time to write. I did see letters from those Nhpum Ga days that you sent to Mom, and your print was about two inches tall. It seemed like when you were calmer, you wrote much smaller."

"Good observation. I was hit in the chest at Wallawbum, and I did not write it down because I would remember it. Back to the May 13th entry, when I wrote that, I thought it was a bullet to my back on my lower left shoulder. I had not seen the doctor yet, but just because you got hit didn't mean you were relieved of duties. It turned out to be shrapnel. Years after the war, it started working its way out, and I had it surgically removed."

"Yes, that makes more sense because I remember seeing a big brown mark on your lower left shoulder where shrapnel must have burned your skin before it entered."

I thought about timing while the Doctor finished with my father's examination. When we left the Doctor's office, I asked about it.

"Dad, if that Jap had been a half a foot closer or your foxhole had been a half a foot closer to the enemy, that bullet would have penetrated your chest," I said.

"Someone once said that every round has a soldier's name on it. I don't really believe that," Dad said. "In the dark of the jungle, it is hard to see in the moving shadows of leaves and branches. It proves to be harder to hit the target, much to our advantage, because I believe we were trained more in marksmanship. When a bullet whistles by your ear or nose, like in the case of Werner Katz, was that luck or God's protection? Most of the men believed that it was God's grace and protection. We didn't talk about it too much because we were always so exhausted. Many did talk with the chaplains about it. In life and death situations, prayers are heartfelt. God listens. Miracles are there for those who listen to God and believe. He hears you. So often, we believe prayer to be the last sip of water before we collapse from dehydration, but it doesn't have to be that way. Just ask."

May 14, [1944]. Sunday. God, please guide and protect us. Carry us thru this safely. Are withdrawing. 1st platoon is rear guard. Rough. God get us thru for Evelyn's sake and in Jesus name, Amen. Marched 8 miles into area at 6 p.m. Chow and bed. Bread and jam.

May 15, [1944]. Mond. Up 5 a.m. ate cooked candy. Took bath. Have bad cold. Chest and head. Monsoon has started. Left out at 9 a.m. Marched 12 slippery and muddy miles. In at dark. Chest cold is bad. Sleep 10 p.m.

May 16, [1944]. Tues. Up Early 5 a.m. Left out 6:30 a.m. Raining. Up muddy hill and down. Pvt Goddard. Covered 11 miles. Rough and slippery. Made candy at noon. Chest is worse. Am the point into bivouac area. Cross 4 rivers. Bed Early. T S my chest.

May 17, [1944]. Wed. Up Early. Move out 7 a.m. Khaki leads. Marched all day. Rained. Good level trails. Marched 12 miles. Swiped 10-1 rations. Swell. Moved out at 7 p.m. Marched another 9 miles. Tired and sore. Close to Mishinau [Myitkyina] our objection [objective]. A BN is there. Sleep 10 p.m.

"I'm not sure if the 16th or 17th was the day that we went on the night march. It appears that it was the 17th because I wrote, 'moved out at 7 p.m.' We had a Kachin guide that was taking us through the hidden trails. When the Kachins established a village, they always made hidden trails for escape routes. These trails are hard-packed and covered with vines that were invisible to those who were not aware of them. When you parted the veil of vines, a whole new trail opened up. The Japanese patrolled the main trails leading into the villages, and we were trying to get through undetected, so we had a guide. We had been warned that we were in King Cobra country, more specifically, the Hamadryad cobra, which is one of the deadliest snakes in the world. It can bite multiple times, and the victim will die within forty-five minutes to an hour. The average length of these snakes can reach twelve feet but have been known to reach eighteen feet and can stand a third of their body length perpendicular to the ground. They can still move forward when they are standing and can strike; therefore, the safe zone can be underestimated. Although they prefer to stay away from people, they can be very aggressive, especially near a nest. We marched in silence during one of the darkest nights that I had ever experienced. It was so dark that it threatened to seep into your pores until you would no longer be visible. Invisibility threatened to separate you from the earth and God, but that is entirely impossible. However, my feet touching the ground and the sound of the other men marching, even though we were trying to be quiet, kept me sane. We were stopped and stood on the trail for a while until the message was passed down that our Kachin guide by the name of Nau had been bitten by a snake. There are always rumors, and the messages don't get passed down correctly. At first, we heard that the Kachin was bitten on the neck by an Asp, which became a redneck [keelback] had bitten him on the ass. We finally got the story straight. He was bitten on the ankle, and it was too dark to determine what type of snake it was. The doctors were called forward and asked Colonel Hunter for permission to use a flashlight. When given permission, they realized that the guide's leg was swelling quickly from what they believed to be the bite of a poisonous snake. Captain Laffin and Lieutenant Dunlap sucked the venom from the wound for two hours. The doctors insisted that if he walked, it would kill him. Colonel Hunter was adamant about continuing to find the hidden trail and sent his horse up so that the guide could ride. During this life-threatening event, we knew little. I never knew that silence and darkness could feel like an elephant on your chest. We would be completely lost if the guide died. Finally, the hidden trail was found, and we continued our march and arrived at the motor road to Myitkyina, which we crossed and made our way to the village of Namkwi. The airfield at Myitkyina was now within five miles."

> *May 18, [1944]. Thursday. Up 4 a.m. fixed packs. moved out at 530 [a.m.]. Held up til 7:30 a.m. Coffee while you wait. Dive bombers raising hell. Troops transports landed. Marched 10 miles into [Myitkyina]. Few Japs held us up. 8 were killed. Got suntan dried and cleaned up. Nips are just a short distance away. Bed Early. God guide and protect us. Amen.*

"We later learned that Colonel Hunter's H force, which was the 1st Battalion's combined Red and White combat teams, along with the Chinese 150th Regiment, attacked the airstrip. When the first Battalion captured the airstrip, Colonel Hunter radioed General Merrill that the strip was secured for air transports and supplies. The Red Combat Team also took the ferry which controlled the crossing of the Irrawaddy River."

> *May 19, [1944]. Fri. Up 5:30 a.m. Rained us out about 7 p.m. Got codeine from doc. Chest is in bad way. Coughing up blood. 500 Nips headed this way from North. No rats for 3 days. God in heaven guide us safely through this and protect us for Evelyn's sake in Jesus'*

name Amen. Got rats late. 5:30 p.m. put outpost 800 yards out. Continue patrols. Bed early. Expect Nips.

May 20, [1944]. Saturday. Rained us out again. Wet and cold. Chinese troops came in on planes. Big battle ahead of us. British crossed Irrawaddy. Nips near outposts. Rained till 11:30 a.m. Washed up. Moved up Road to another position at 1 p.m. Sun is hot. Dug in BAR 100 yards apart. Ate. Bed Early lots of mosquitos.

May 21, [1944]. Sunday. Myitkyina Up early. Packed up. Rained like hell. Attacked down road to Myitkyina. Got hit by nips at road junction. Bullets everywhere. God guide us safely thru this & protect us in Jesus' name Amen & for Evelyn's sake amen. Fired all day. Rained. Wet and sore. Pulled back and built perimeter. 75's hit nips. Lost mg was bayoneted to death. Took mg.

"The Nip attacks came from the north. We had left Charpate, which was about five miles from Myitkyina so that we could take the road junction north of Radhapur; however, we encountered the nips who kept us pinned down. We dug in."

May 22, [1944]. Monday. Up 4 a.m. rough last night. Mosquitos. Bullets, artillery, mortar. Dug in deep. No food again 3 days before. Rolled packs. Raining. Got food At 930 [a.m.]. Moved out 10 a.m. Set up old position in village. Have a hell of a headache. Got 15 nips on trail. Made candy. Raining. Natives nice. Bed early. Sweating out. Coughing up blood. Japs firing all night.

"I have never felt more desperate in my life. Enemies without and within. Something bigger than myself was desperate to take me out, which made me more determined to survive. It's hard to have a good attitude when you are deathly sick. I just wanted to sleep. I needed medical help and vowed to go to the doc the next day," Dad said.

>*May 23, [1944]. Tuesday. Rained all night. Men's nerves are breaking. Trigger happy. 14 men being evacuated. Typhus. Cold and wet. No news. Sicker than hell. 102.8 fever. Typhus. Slept all evening. Saw A BN. Bed Early. Same place.*

"The overwhelming numbers of Japs, around 5,000 of them, pressured us to the point that Colonel Hunter ordered Colonel Beach to disengage contact with them. We went to the railroad, and the Japs overtook Charpate. This was problematic because the Japs now occupied two of the towns that were on the routes to Myitkyina. They tried to hit us from the rear, but we stood our ground and fought off the attack with the help of our artillery. Our 3rd Battalion was depleted because of sickness, yet we fought on. Night fighting was the worst. We had to stay in our foxholes. If you got out or stood up, your own men wouldn't be able to see you and would assume you were the enemy. When there wasn't a firefight, and we had to get out of the foxhole, we had a code word so no one would fire on us. I remember one code word was 'lollypop.' The Japanese had a hard time pronouncing Ls.

"On this day, our worst nightmare happened. The Japs broke through our perimeter during the night and killed so many of our men and officers. I don't know how I survived it. My survival was so incredible that I can only give credit to divine intervention."

>*May 24, [1944]. Wednesday. Up all night. Nips hit us last night and raised hell 920 [p.m.]. Inside of*

perimeter. Killed Lt. Smith and Hogan, Sgt Mattlock, Shafer, and Burton. Bayonetted and shot 7 more men. No sleep fired all night. Nips all around. Rain continues. Moved out at 11 a.m. Hit Mogong [Mogaung] Rail Head. Set up perimeter. I was given MET slip and went to airstrip. Ate then. Was picked up by troop carrier. 1 hour and 15 minutes to Ledo. Slept in ward. Is wonderful. No nips Close. Have blue spots over me. Spitting up blood and shrapnel wound.

Earlier, I had gone to the doc because I suspected my fever was high, and he gave me a ticket out. I put my medical evacuation tag next to my diary and Evelyn's folded letter in my shirt pocket and ran five miles through the jungle looking out for Japs, trip wires, and booby traps. It was a wild run. Sometimes, I had to stop to cough. I tried to cover my mouth to suppress the noise because I didn't want to be detected. My nerves were fragmented when I reached the cleared area around the Myitkyina airstrip, which we held, but I still had to run across that clearing, dodging sniper bullets to get to the medical tent near the plane. I ran with my backpack and my M1 with 103 degrees of fever, a shrapnel wound in my groin that continued to bleed, and a shrapnel or bullet wound to my back. I wasn't sure. I was determined to be evacuated. I ran in ankle-deep mud, skirting around bomb craters filled with water, and crossed a large, flat, grassy but muddy clearing where crashed planes and shelling debris lay pushed to the side so the planes could come in. I finally got to the airstrip. I sat on the ground and caught my breath outside of the medical tent. I coughed my head off and spat blood. I couldn't believe I made it out of that wilderness of death. When I calmed down and stopped coughing, I checked in at the medical tent and then ate my K rations. That big grey and olive-drab Gooney bird [C47] was my lifeline to help me leave the fighting. I couldn't believe I had made it out and hoped I had left Burma forever. I fought my way out every step of the way. The fuselage of the plane was lined with stretchers, and medics were there attending to the wounded. We were all wounded or sick. I sat on a bench and looked out of the window. I thought of the last narrow escape and of the rounds that killed Lieutenants Smith and Hogan, Sergeants Shafer, and Burton, and my friend Sergeant Tom Mattlock. I later heard that Lieutenant Smith was in a foxhole on the telephone when he was shot at close range. Although I was saddened by any of my fellow Marauders that died, I was so grateful that my squad and I had not seen eye-to-eye with Sergeant Mulligan, who reported to Lieutenant Smith, because we would have never been moved to the first platoon and possibly died when the Japs broke into their perimeter.

"I have never forgotten my friend Mattlock and seeing him run down the trail with grenades exploding and hearing that machine gun that shot him in the back. It felt like that machine gun had hit my own back. The Gooney bird took off, and I felt the lift into the air. I sat next to a window and watched where we had taken painstaking steps to clear the Nips.

"I looked at that God-forsaken land where my brothers-in-arms died and were buried. The destruction could be seen for miles. I had no doubt that in a few years, the jungle would win in the end, recapture the destruction, and reclaim the Ledo Road. It was an amazing feeling to sit on a plane with no Japs to watch out for. No tripwires. No cobras, no kraits. No shelling and grenade fragments to look out for except for the one I carried in my groin. I looked at the soldier across from me. Our eyes connected, and we both swallowed hard and looked down to keep from crying. However, I would have loved to have been able to cry. I could not. My emotions were stunned. I had to pinch myself to know that I was alive. I didn't feel anything. My heart felt like it had been wrung out like a dishrag of all the blood it pumped. It felt limp and exhausted. We were one step closer to home. My body was exhausted and broken, and I worried that every time I coughed up blood, I had more battles to conquer. I still had to practice bravery."

May 25, [1944]. Thurs. Up early Swell breakfast. Maner and I caught ambulance to Marguerite hospital [20th General Hospital in Margherita, Assam India] Ward. B17. Took swell shower. Went to the red cross toilet then back to bed. Doc looked me over. Looks bad. No mail, nothing. Still spitting up blood. Black spots. Bed early.

I had forgotten how comforting it was to have the luxury of a shower, a toilet, and a bed with sheets. I didn't have to worry about sleeping against logs that may have hidden snakes and poisonous insects. I didn't have to worry about Japs trying to kill me.

May 26, [1944]. Friday Up at 7:30 a.m. Raining. Still in Margherita General Hospital 4 docs came around. Looked me over. Bad. Feel worse. Good chow. 3 more Doc's came in to look me over. Bad case. Wrote Evelyn. Good chow. Bed. 930 [p.m.].

May 27, [1944]. Up 7 a.m. good breakfast. 2 Docs this morn. Read books. Not much else to do. Tired of laying around. Feel bad. Bed Early won 25 Rupees.

"While I was in the 20th General Hospital, news came to us from the incoming patients that my friend and former squad member Al Mahmood was shot in the abdomen. He was in an open field trying to knock out a light machine gun when he was hit. The bullet went through his large and small intestine and out through his tailbone. He turned around and walked 50 or 60 yards back where he received aid. He had surgery and long-term treatment. Many years later, I heard that he survived. I was so glad."[37]

Mary Evelyn 1944

"Mom," I said, "What do you remember? How did you know he had been evacuated?"

"Well, let me tell you the story," she began. "By 1944, my family and I were living in Lake Charles. I had returned from Fontenot's grocery store on the corner of Kirkman and Sycamore and just finished putting the groceries away. I looked through my rations book and dreamed of the day the War would be over and foods like sugar, coffee, and lard would be freely available — especially coffee. My daddy, your grandfather, used a tightly knitted, new sock and made a filter out of it for his coffee, so the water stayed in the sock longer to increase the strength of the brew so that fewer coffee grounds were used.

"We always got meat from Granny and Grandpa when they butchered a cow or hog. Daddy and some of the neighbors would help Grandpa butcher. He had a smokehouse in the back, and he smoked much of the meat. This was before freezers became popular and made available for regular folks to buy. I know many people packed their meat in a barrel of lard to keep the air from getting to it and causing it to spoil. Mama always had a vegetable garden, so our pantry shelves were lined with jars of squash, corn, pole beans, and other vegetables from Mama's garden that she had preserved.

We made preserves from the figs from our own tree. We sweetened everything with honey from Daddy's beehives because we couldn't get enough sugar stamps. Sometimes, Aunt Sarah would give us her sugar stamps because she rarely used them.

"We treated those ration books like a prized possession. Mama had an old ration book that she kept as a keepsake. She read the ration statement out loud. The War Ration Booklet Three stated that 'punishments ranging as high as ten years' imprisonment or $10,000 fine or both may be imposed under United States Statutes for violations thereof arising out of infractions of Rationing Orders and Regulations.' We took the threats seriously. On the inside of the front cover, it states: 'Rationing is a vital part of your country's war effort. Any attempt to violate the rule is an effort to deny someone his share and will create hardship and help the enemy. This book is your government's assurance of your right to buy your fair share of certain goods made scarce by War: Price ceilings have also been established for your protection. Dealers must post these prices conspicuously. Don't pay more. Give your whole support to rationing and thereby conserve our vital goods. Be guided by the rule: 'If you don't need it, DON'T BUY IT.'"[38]

"I remember putting the ration book in a cabinet drawer where we wouldn't lose it. I began to tidy up the kitchen on automatic before we began to cook for dinner. I made sudsy water and washed the dishes as I looked out the window. I dreamed of the day Larry would come home to me, and we would have our own house and my own kitchen. Somewhere between my dreams and the iridescent bubbles, reality struck.

"Our dog Chippy was curled in sleep outside the screen door. Daddy was in his garage changing the oil in the car. Mama was hoeing in her garden. Chippy started barking. I hushed him when I saw a man coming up the walk with a telegram in his hand. My heart skipped a beat because telegrams usually told of homecomings and times of arrival, and I thought maybe Larry was coming home. The delivery man came to the door and handed the telegram to me. I stared at it with such anticipation and ripped the envelope open. When I read the message, I couldn't believe what I was reading. It was from the War Department. Larry was MIA [Missing in Action]. I fell to my knees and screamed so loud I thought the whole neighborhood would hear me. Daddy and Mama came running in to find me in a heap on the floor, sobbing. I had so much sorrow; I couldn't take it. My family and I prayed and prayed and prayed. I went to bed for a week. I was useless. About a week later, I received another telegram. When the man delivered it to me, I took it and held it cautiously. It trembled in my hand like it was my own heart threatening to die from the potential news.

"It was the good news that Larry was found and was in the 20th General Hospital in Margarita, Assam, India. The bad news was that he was diagnosed with scrub typhus. Although the typhus diagnosis was deadly, he was alive. I remember Mama and Daddy hovering around me. I screamed, 'he is alive!' and I broke down with tears of joy. He didn't go through the horror of fighting in the jungle to die now. He would live, I told myself. I, too, felt resurrected and was called from the tomb."

Mama lifted her head and wiped her tears. I hugged her and wiped mine. I felt incredible sorrow for all the wives and families who learned of the death of one of their loved ones and what the American families went through to preserve our nation.

20th GENERAL HOSPITAL

MARGHERITA, ASSAM, INDIA 1944

May 28. [1944]. Up early. Made bed. Another blood test. 3 times they have stuck me. Rough. Good meals. Sun is shining swell. 10 new cgt, socks, and drawers. No shoes. Won 31 Rupees. Bed late. Saw lot of old friends. God guide us safely and protect us on our way for Evelyn's sake, in Jesus' name. Amen.

"Dad, were you smoking in the hospital? You were coughing up blood. What were you thinking?"

"In 1944, we did not have all the research on the effects of cigarette smoke like we have today," he said. "When I was in India, everyone smoked, even the doctors. I quit when I returned to the States by the recommendation of the doctors."

May 29, [1944]. Monday. Up early still nothing to do. Laying around. It's rough. Wrote to Evelyn, Atchinson, Ida. Tired as hell. Read 3 books good chow Won 23 Rupees poker.

May 30, [1944]. Tuesday. Up early had to go before over 100 docs for strange new case. Don't know what it is. Tried to get barracks bag. Lost 3 Rupees poker. God carry us thru safely and get us back to the States safely for Evelyn's sake in Jesus' name. Amen.

"Doctors from London and many more came in to examine me. They thought that I had a new form of the plague. Around the year 1350, there were three types of plague: Septicemic, Bubonic, and Pneumonic. They were commonly known as the Black Death. My symptoms were similar to those of Septicemic plague, which was the least common of the three plagues but the type that gets into your bloodstream. It began with a severe headache and high fever. I had enlarged lymph nodes, coughing up blood, black spots all over my body, and a high fever. My organs were enlarged, which is known as visceromegaly.[39] All those doctors were trying to make a diagnosis. As time went by, it turned out that I was in the end stages of Orientia Tsutsugamushi fever, which is caused by larval mites. With antibiotics, good hospital care, blood transfusions, and by the grace of God, I survived."

"Praise God," I said. "It was a shame that the Marauders did not have penicillin available to them then. It was the wonder drug of the 20th century. Alexander Fleming of St. Mary Hospital in London discovered it by accident in 1928. It was tested for military use in the spring of 1943.[40] For God's sake, why couldn't your outfit have had penicillin in their small arsenal of drugs? I've been doing some reading," I said, pulling out my notes, "The major drugs the docs used for the Marauders were Crystalline Sulfanilamide, a yellow powder placed on open wounds; Sulfadiazine tablets, and Atabrine. Sulfadiazine was used to treat malaria, urinary tract infections, and an infection from toxoplasma parasites. Atabrine was the alternative drug for quinine called quinacrine, which was also used to treat malaria.[41] However, in the field, it was known as Atabrine. If only penicillin had been available to your unit, so many lives would have been saved! The allied soldiers that stormed Normandy Beach on June 6, 1944, had penicillin available to them!"[42]

I blurted out in exasperation. "They could have made it available in March and April of 1944. This was a failure of some adviser for the Marauders. I know the hospitals in India had limited use of penicillin. Did you know if you were given penicillin?

"I don't know. So many died from scrub typhus it makes me think that they did not have penicillin. At any rate, it sounds like you did your homework," Dad laughed, impressed.

"Yes, I did."

"So often, the weather prevented the planes from flying and landing, causing treatment to be delayed," Dad explained.

"But if a Mobile Surgical Portable Operating Unit had been attached to the Marauders, many lives would have been saved. These American surgical units were used in Burma but were attached to the Chinese, which were under General Stilwell's command.[43] Why didn't they attach one to your unit?! It doesn't make sense," I protested.

"It's okay, Windy Lindy," Dad said, trying to calm my temper. "I survived. I am where I am supposed to be. You would not be here if I had not survived," he pointed out.

I nodded at the magnitude of that statement.

"My children would not have been here as well, and their children's future children. All that possibility of life that so many men who died over there never realized," I said.

Dad paused a moment and let what I said seep in. His silence honored the brevity of my words. Then, he took up the gauntlet and continued his story.

May 31, [1944]. Wednesday. Up early feel like hell.

Docs looked me over. Young Brought in 30 rupees worth of candy and other things. Won 8 rupees in card game. Is hot as hell. Bed early. Letters to Mama.

"After Colonel Hunter's H Force, which incorporated the 1st Barralion, had taken the Airstrip at Myitkyina, I heard that General Stillwell insisted on finishing the mission by taking the town of Myitkyina. So many of the men were in the hospital in India, and of those, around two hundred were ordered back to the battlefield by General Stillwell."

June 1, [1944]. Thurs. Up early. Got haircut. Had watch fixed. Good meals. Hot as hell. Wrote Evelyn. Sent Jap & Chinese money. Bed late. Took shower. Saw Westy & Doeling.

"Sadly, on this day, I received the news that Colonel Henry Kinnison of K Force had passed away from scrub typhus on May 31."

June 2, [1944] Fri Up early good breakfast. Went to get xray. Saw Joe Diskin Good boy. Played poker. Lost 22 Rupees. Bed early. Read books.

June 3, [1944] Sat. Up & at em. Feel Better this morn[ing] 211 of 2 Bn is now out of Myitkina [Myitkyina] good. Saw Westinghouse. Hot. I took xray. Saw Joe Diskin is going into Myitkyina.

June 4, [1944] Sund. Up early not much doing. Good meals got my clothes. Won 45 Rupees. Saw same old friends. Bed early. No letters. Wrote Evelyn & others.

June 4, 1944

My Darling Evelyn,

A few lines honey *to let you know that I'm alright and I do hope my precious you can say the same for yourself even though I haven't heard from you to know it. I'm sure that God will take care of you for me for I do know that he has passed miracles time and time again to get me out of the holes or traps I have been in. Only he in answering our prayers could have gotten us through.*

Please give my most loving regards to all. Am very sorry that I can't give them to all in person, but just be patient and I don't believe it will be too long for God is with us. I know this sounds like a preacher but if you only knew how he has pulled us through you would think differently. I miss you so much my darling, I don't know what to do. My thoughts are continually of you.

With all my love,

Larry

"I thought I was getting well, but that's not the case. On June 19, I began to relapse."

June 19, [1944] Feel pretty bad today. Fever 101

Blue spots. It's a relapse. Lots of docs.

June 20 [1944] Worse than ever. F. 102.6 Spitting up blood. Gosh feel rougher than hell 9 letters from Evelyn. Very Sweet.

June 21 [1944] Worse Still more blood & f. 103.6 getting worse. God please help me in Jesus name Amen.

June 22 [1944] God I'm getting worse. More blood spitting up F 104. Will get blood trans tomorrow. Weak as hell.

June 23 [1944] Had blood trans today 1150 cc of it. Fever 105.6. Ice packs all day. Feel rough. No more spots. No more blood.

June 24 [1944] Feel better today. Fever lower. No blood come up. Ice packs okay I guess.

June 25 [1044] Feel better today. No fever. No blood up. Can't eat & very weak.

June 26, [1944] Am snapping out of it friends came in Not much to eat. Still weak as the Devil. Letters from Evelyn Swell to hear from her. God is kind.

June 27, [1944] Wrote Evelyn & Mama today. Feel a lot better. God please carry me back to States soon & safely for Evelyn's sake in Jesus name. Amen

"It's funny, but when I was fighting the Japs, I did everything in my power to survive using all the lessons and skills I had learned in training or by experience, plus my prayers beseeching God to save me. Now, I couldn't see the enemy, only its aftermath. I was absolutely in the arms of God and prayed for good doctors and nurses to use their skills to help me survive. I had so many doctors looking at me as an anomaly that they took a personal interest in my survival. I called it my summer of surrender—not to the Japs, but to the Almighty.

I left the Myitkyina airstrip on May 24, 1944, and stayed in the hospital in Margarita, India, which I later learned was the 20th General Hospital, until July 28, 1944. During that convalescent time, I relapsed, and my fever soared to 106 degrees. They packed me in ice more than once. One of the clerks from headquarters wrote letters to Evelyn for me. All I could say was to tell her that 'we were still licking the Japs,' and that 'I love her.' I saw guys come in, heal, and they were back in battle. I heard of those who died from Typhus fever. Father Barrett was one of them. That good priest who gave us emotional support, buried our dead, and suffered along with us was truly a witness.

"On July 28, 1944, twenty-one of us flew from Ledo then to Goya on a C-47. We changed to a C45 and flew to Agra. The next day, we flew to Karachi. While I waited for the next flight, I managed to buy a moonstone to be made into a ring for Evelyn. The moon was so memorable to us. The jeweler that I bought the stone from said that the moonstone was a special stone because it promotes a calm mood and alleviates fear associated with change. I thought it was funny because change was the least of my fears. In fact, my fears had been tested to the max. I wondered if I had passed the test or learned the lesson that I needed to learn. The one thing I learned was not to volunteer for a 'hazardous and dangerous mission' unless you wanted a nightmare of memories that shredded your psyche like razor blades. Left Karachi on a C-54 and arrived in Arabia. Then, on to Egypt near the Suez Canal. Flew all night to Africa's Gold Coast. On August 3, 1944, we flew to Ascension Island and then on to Natal. On August 4, 1944, we flew to British Guiana. I was so sick on the plane that I couldn't eat. Landed in Puerto Rico at 9 p.m. I was sick as hell. Finally, we arrived at Miami Beach. I was too sick to enjoy the fact that we had landed on the mainland of the United States of America. The country of my birth, the country, and its people that I fought for. Even though I was so sick and weak, I stooped down, got on my knees, kissed the pavement, and privately thanked God for bringing me home. One of my buddies helped me up. They checked my temperature, and we went through customs. It was strange to be back in American society. People were going about their comfortable jobs and making mountains out of molehills. They had no idea of what we had been through. A bus was provided for us and took us to the Biltmore Hotel, which served as a hospital during WWII. I ate ice cream and drank *Coca-Colas*; such simple pleasures meant so much."

I drifted into my own memories of the time in the train station all those years ago, when I was so sassy in my assessments of those soldiers that gave us grief while we waited for the train to go to Arizona. I had no idea of what they had gone through. I decided to try to look at all perspectives before concluding about other people's lives and motives. I wiped a tear from my eyes as Daddy continued his story uninterrupted. I let the tape roll.

"The greatest gift was to hear my darling's voice over the phone. I began calling at 7:30 and got through at 10:30. Boy, oh boy, it was so swell. I got choked up when I first heard her voice. She knew it was me because the operator asked if she would accept a collect call from me. When she said, 'Larry', I slid down to the floor of the phone booth, covered the receiver with my hand, and cried. It was as if there had been a dammed river, a cesspool of horridness that had built up inside of me. I thought I couldn't ever cry again. With a surge of adrenaline, I broke the dam, and it drained me until I was so weak that I couldn't talk to her. I couldn't believe my ears: the love of my life, my wife, the one who gave me the courage to come home, said my name with such anticipation in her voice. I took a deep breath and said, 'My darling, I've missed you so much.' She broke down in sobs. She had no idea of what I had been through. I have protected her from the knowledge of that gruesome War. I also talked with Mama and Papa, Frankie, and Oscar. Mama cried. It was so good to hear their voices. I stayed in the hospital in Miami until August 2 and then flew to Tallahassee, Florida, then to Jackson, Mississippi. From there, we flew over to Lake Charles, Louisiana, at 3:15 p.m., never knowing that I would make my living and raise our family in that town. I strained to look out the window as we flew over and wondered if Evelyn had looked up, searching the sky for our plane. We landed in Tyler, Texas, and then on to Dallas. From there, we went to San Antonio, Texas, and then to Santa Fe, New Mexico, where my fellow Marauders and I were admitted into the Bruns Army Hospital.

"Evelyn and her sister Ethel took a train to Santa Fe. They had never been out of the state of Louisiana and had never been on a train, but love and desperation make you brave. Santa Fe was hot in August, but there was always a breeze that made the heat tolerable. I like to think of the weather's cooperation as if it was a part of the great choreography of the reunion of two young people whose love and devotion defied the odds of a world at War. The brilliant blue sky hovered over Evelyn and Ethel as they scurried to get a cab to the La Fonda Hotel on the Plaza. Evelyn was so intent on our reunion; it wasn't until later that she appreciated the distinctive Pueblo Revival Architecture. Evelyn later told me that she had dreamed of running to me at the train station wearing the yellow dress she wore when we first met. She intended to throw her arms around me and kiss me, and I was to hold her so tight that nothing could pull us apart. It didn't happen exactly like that.

"I was in the hospital with heart damage from scrub typhus fever. I didn't want her to see me in a hospital bed, so I arranged to meet them at the hotel at 108 San Francisco Street. It occurred to me that when I first left her and the States for War, it was from San Francisco. Now, I am meeting her for the first time since I returned from the War on San Francisco Street. God or Saint Frances had to have a hand in this. Friday, August 25, 1944, I was up and dressed and waiting for her call. At last, she called the Bruns General Hospital. 'She's here! She's here!' I shouted to everyone in the lobby who knew our story. They started clapping. I got a pass, and a friend dropped me off at the plaza. I bought flowers from a street vendor and walked into the LaFonda Hotel. I had dreamed of walking down San Francisco St. in my Army uniform, looking dapper, and Evelyn would be standing outside of the door of the LaFonda. When she saw me, she would run to me down the middle of the street, stopping traffic. She would throw her arms around me, and we would kiss in the middle of the street. The people in the cars would get out and clap for the touching reunion. It would be so romantic, and you know, movieish. It didn't happen like either of us had imagined.

"I went to the lobby at the LaFonda with its grand beamed ceiling and old Mexico charm. When I entered, I hesitated and realized how very self-conscious I felt. I was skeletally thin as a concentration camp survivor. I don't think I mentioned to Evelyn that I had lost fifty pounds, although I had gained some back since

I had good food in the hospital. Maybe she won't want me anymore. Maybe life with me was not what she bargained for. Maybe she would change her mind when she saw me. I went to the counter and told the receptionist my name and that my wife was staying there. 'Oh, we know, sir,' she said brightly as if she were in on a big secret. 'Your wife told us all about it—that you are home from the War, and this is the first time you have seen each other for a year. By the way, welcome home, sir.' Evelyn later told me that after she had received the message from the front desk, she and Ethel ran down the hall of the hotel. When she got to the entrance of the lobby, she stopped, combed her hair with her fingers, straightened her canary-yellow dress, and took a deep breath. I looked in the grand mirror behind the receptionist counter and saw her enter the room. She hesitated and looked around. She didn't recognize me, I thought. I turned slowly around, and she gasped. She was going to throw herself into my arms, but she stopped, shocked at my thin appearance. She turned her head sweetly and came to me gently, and we fell into each other's arms, weeping and weeping. She said, 'Thank God for He answered my prayers.' The owner of the hotel had a waiter bring complimentary champagne, and he and the staff gathered around with big smiles and clapped and cheered for us. We would begin again.

"On August 25, 1944, I saw Evelyn for the first time since I left Burma. On September 7, 1944, I was discharged not only from the hospital but from the United States Army. My beloved wife Evelyn was there to help me; together, we entered a new era of our lives."

Linda S. Cunningham 2022

Years later, at a Merrill's Marauders reunion, Dad heard the fate of the horses and mules and his favorite horse, Red. The animals were under the jurisdiction of the 252nd Quarter Master Remount Squadron, which was re-designated the 475th Quarter Master Remount Troop. The horses and mules were turned out into corrals and green pastures, where they were rehabilitated and gained weight. However, there were some that were diagnosed by the Veterinarians with chronic diseases like surra, negana, swamp fever, and other disabilities. The Foreign Liquidation Commission was given the authority to dispose of the horses, mules, and equipment. Some of the animals were sold, and according to J.D., Red was sold to a local man for his son. Dad was elated when he heard the news that Red survived. In his mind, he thought of the little boy with big brown eyes who marched next to his platoon when he first arrived. Maybe that was the man that bought Red for that boy. However, the chances are that it wasn't the man at all, but the thought of it created order and closure in the chaos of that period of time.

I read an article in 2008 in the *Burman News* by Edward A. Rock, Sr., that some of those animals had to be put down for humane reasons. He wrote, "Thus, on a cloudy, gloomy, very sad January morning, approximately 125 horses and 75 mules were carefully and slowly led on a "Last Round-Up" to a trench on the far side of the depot, which was used as a burial grave. The disabled or diseased animals were gently led to the edge of this grave, where each animal was shot in the head by a veterinary officer. Those of us who were there watched, not speaking (knowing it had to be done for humane reasons) with heavy hearts, tears in our eyes, lumps in our throats, and feeling that old friends and comrades who had shared our trials of the jungles, monsoons, malaria, and enemy fire in battle were gone. As personnel sergeant and a sentimentalist, that evening before sunset, I request[ed] the officer-of-the-day- to march a rifle squad to the burial site. Complying with the request, a finale farewell volley was fired in salute to those brave, courageous animals that were a proud part of the China-Burma-India Theater of Operations history."[44]

-*-

Shortly after Dad and I finished transcribing his diary, we wrote to the Pentagon requesting his medals. I drove to Lake Charles the day he received them.

"This wouldn't be happening if you hadn't transcribed my diary and contacted the Pentagon," Dad said. "I thought they would put me in the stockade. Non-coms are not supposed to keep a diary, you know. It could fall into enemy hands."

Dad stuck the point of his Swiss Army knife into the edge of the box and slit the tape. He removed the resistant top of the box. In individual plastic bags, he took out medal after medal: Bronze Star, Presidential Unit Citation, Purple Heart—inscriptions that read, "honor, fidelity, efficiency," "freedom from fear and want," "freedom of speech and religion." Others read, "WWII service during the limited emergency proclamation by the president" or "during the unlimited emergency proclamation by the president...'", "Asiatic Pacific campaign" and the Merrill's Marauders' insignia. The Distinguished Unit Citation awarded to the unit described the campaign thusly, "After a series of successful engagements in the Hukawng and Mogaung Valleys of North Burma, in March and April 1944, the unit was called on to lead a march over jungle trails through extremely difficult mountain terrain against stubborn resistance in a surprise attack on Myitkyina. The unit proved equal to its task and, after a brilliant operation on May 17, 1944, seized the airfield at Myitkyina, an objective of great tactical importance in the campaign, and assisted in the capture of the town of Myitkyina on August 3, 1944."[45]

Daddy's moment was long overdue, I thought.

He sat quietly, staring at each medal, running his fingers over those symbols of bravery. So many of those years, Daddy had never accepted disability payments from the government. In the honorable ranger tradition, he said that he "was only doing his duty." Fifty-two years after the War, the United States Government was honoring this old soldier, and he was so proud. There should have been a band, a parade, and a general pinning those medals on his chest. But it was just me, sitting on the floor at his feet, and Mama sitting in her chair. Daddy stared into the distance, engulfed in the incredible silence of a war too big for words—a silence broken only by the squeak of the rocker.

-*-

I treasure the conversations that I had with my Dad when I transcribed his diary. Although my father was always a hero in my eyes, I realized that he truly was a war hero, as were all the Marauders. My Dad came back humbled but extremely grateful and wondered why he deserved to survive when so many of his brothers-in-arms did not. He had his problems with PTSD (Post Traumatic Stress Disorder) that made me so curious as a child.

He would scream in his sleep and say things like, "Get the gun, get the gun!" On many occasions when airplanes flew over, Mama saw him dive face down in a parking lot, covering his head with his arms. My brother witnessed him diving on the street, skinning his nose and arms when a car backfired. There was no psychological help in the small town of Lake Charles, Louisiana, in the forties and fifties, even if he could afford one. He had to tough it out.

Yes, on August 10, 1944, the 5307th Composite Unit (Provisional) was disbanded and became nonexistent, but the aftermath of that experience existed in the minds of the surviving Marauders and the home-front wounded—their families–for their entire lives. Merrill's Marauders will live in our consciousness if we as individual citizens honor the patriots of America's history for their loyalty and love of country.

It is the intention that counts. Their intention was to preserve the goodness of America when evil rose through the ranks to become a Hirohito, Emperor of Japan—the bomber of Pearl Harbor and the massacrist of Nanjing, or a Hitler who became the annihilator of eleven million Jews and sympathizers. Our history is the anchor of time that brings order and retrospect to our lives. Time is a trickster. It makes you look old but feel young in your mind. It brings joy and sorrow in its fists to become your choice. The two most important days in your life are the day you were born into time and, most significantly, the day you exit from time. The in-between is where you live the 'why' of your life.

I remember the lyrics of a little song Mama sang to me as a child, "This little light of mine, I'm gonna let it shine." I like to think of life as being born a candle in the wind for us to protect, but for Dad, it was a candle in a hurricane. It is our choice to join the light of other candles to enlighten the world or to lay your candle down and start an inferno. Hell is only a choice away. It is our choice to choose between good and evil.

I often regret not knowing my father's story when I was younger so that I would have been more understanding of his mood swings, his brooding, and his need to be quiet and alone. My father was of the generation that pulled themselves up by their own bootstraps. He set to work and did not waste time: those ticks of the clock, those calendar days. He wanted to make something of himself— to be industrious, to understand the goodness of God that he was spared and to bring enlightenment to the citizens of our country on the sacrifices that constitute a true patriot.

Dad went back to school because he knew he was not physically able to do a lot of manual labor. He graduated from the University of Arizona and established his own Accounting and Tax Consulting business, which he maintained for 42 years in Lake Charles, Louisiana. He served in numerous organizations and clubs in Lake Charles, such as the Jaycees, Boys Village, The Optimist Club, the Chamber of Commerce, and Potentate of the Habibi Temple (Shriners), to name a few. He also served as Director of Civil Defense for Calcasieu Parish and the President of the National Civil Defense Counsel. He was proud to be one of the Marauders and served on the board of directors. He led a very busy life, partly, I think, to not only avoid the memories but to do something worthwhile for his miraculous survival.

The Marauders had a special bond of never leaving anyone behind. When Dad received the shrapnel wound to his groin, which came so close to ending his progeny forever, he was crawling to help one of his buddies who had been hit. He told me, "I eased out of my foxhole and was crawling on my belly to reach one of the guys that had been hit and who was screaming for help. At first, I believed it was shrapnel from a grenade explosion because they make flat trajectories. It happened so fast, but as I thought about it,

I realized that it was a tree burst, where mortar hits a tree. Usually, that causes lead to rain down on you. Within seconds after it hit, a second mortar came in. The thermal heat blasted me and threw me twisting in space. Then, the shrapnel from the encasement shell exploded and dispersed everywhere, hitting trees and ricocheting. Next came the tree splinters. When the shrapnel fragmented and flew everywhere, I got it in the groin. When I came to, I found myself in a different location. The shrapnel wound burned like hell. I used a Japanese flag with its big red circle that I had been carrying in my pocket to staunch the blood until I could get to the aid station."

The 75th Army Ranger Regiment carries on the tradition of 'no one left behind'. They proudly wear the Marauder Patch that was designed by 2nd Lieutenant Charlton Ogburn and other Marauders. The Burma star references the location of the action, the sun from the Chinese National flag refers to the coalition soldiers who were instrumental in their success, and the lightning bolt refers to the speed and mobility of shock troops that their unit code name Galahad infers.

"Why is there war, Dad?" I asked out of anguish from hearing his story.

"I suppose it starts from greed and the need for power."

"This is what I think. During the Stone Age, the species was trying to survive. So, they didn't fight among themselves that much. Then, during the hunter/gather era, no one owned the land that they hunted on. They were still trying to survive. After the nomadic period, people settled in one place and began claiming land and producing crops. Because of the surplus of crops, 'yours' and 'mine' became prevalent. Then the 'have nots' stole from the 'haves,' and from there, *conflicts escalated*."

"That makes sense," Dad said. "I think we are both right."

-*-

The other component in this memoir was the belief in miracles— miracles that defy natural law. I asked him about the miracles, and he said, "War is a chaotic and fierce confrontation with your own mortality. As they say, 'There is no atheist in a fox hole.'" Dad experienced the miracle of being hit in the chest area near his heart. The round did not penetrate but severely bruised him. He later found the slug in his pocket. He kept that slug as a reminder of the miracle he experienced. "That was God protecting me," he told me. In fact, his entire squad survived.

I remember saying to him, "Dad, do you remember when the doctor at M.D. Anderson suspected metastatic breast cancer ten years after chemotherapy, which was very unusual?"

He nodded yes. "I went out and sat on my tractor and put my

hands over my face and cried," he said.

"Awe, that is so sweet. I also experienced a miracle," I said.

"I thought so as well."

"During that time, I was going to M.D. Anderson for my check-up every year. On the day I was to go, I was in church that morning and trembled from nervousness. As I walked out, the priest came up to me and asked me what was wrong. He was new to our parish, but I asked why he asked. He said, 'I see it in your eyes'. Then, I explained that I was going to M.D. Anderson for my test results. He said that he would go with me, and he did. When I met with my doctor, he immediately said, 'Bad news. You have two hot spots on your spine and a lump (which was also palpable and visible) in your other breast.' After discussing further tests, I left devastated. After ten years, I thought I was home-free. I had to wait a week before I went back for the additional tests. During that week, I continued going to Mass every day.

"One morning, I got there early to have some quiet time. The church lights had not been turned on. As I knelt, a presence came to me. It was nothing that I saw with my eyes, but I knew it was Christ standing before me. If you were blindfolded and sat quietly in a room, you would know if a person was sitting next to you. It was that type of feeling. He asked me through telepathy, 'Do you want to be healed?' I said, 'Lord, I know you love me, and you want what's best for me— whatever you want to do with me, it's okay.' He asked me again, 'Do you want to be healed?' I said again, 'Lord, I know you love me— whatever you want to do, it's okay.' Finally, he asked me for the third time, 'Do you want to be healed?' and I said, 'Lord, it's okay. I trust you.' I sat there in wonder that my small being of nothingness could be so filled with His loving presence. After three days, the lump went away. I called the doctor and told him it was gone. He said to come in anyway as scheduled. During this waiting period, the priest had arranged a healing Mass for me and other parishioners. I went to the service but did not go up to the altar to receive a blessing. The priest asked later why I didn't go up to receive the prayers and the blessing of the sick, and I said, 'Because I am not sick.' I went back to M.D. Anderson for the ultrasound biopsy, and the tech worked on me for several minutes and couldn't find anything. She left and returned with the doctor, and he checked as well. There was nothing there.

The MRI results were the same; everything was normal, no cancer. My doctor put in my chart, 'very puzzling.'"

Dad sat there in wonder.

"Dad, do you realize the miracles have followed us?"

"Yes," he nodded. "I had more miracles than one. Looking back, I realized that I was being protected."

"Dad, I said, "The cancer and chemotherapy I had was the best thing that ever happened to me because I would have never been this close to Jesus. I always told Pat, next to Jesus, you are my best friend. He helped me through it."

Dad nodded quietly.

"My wounds and sickness were a painful reality, but the experience of my mind was just as real as matter and festered like an open wound," he began and cleared his throat. "The rifle I fired; the Gurkha knife I used to slice throats—I thought I was a killer. I told myself that to bolster my sometimes-flailing courage. A killer was not who I was. It was survival at the core. It weighed heavily on my soul. I look back at it, and I realize my Evelyn and the horrors of fighting in the jungle with Merrill's Marauders were the best things that happened to me, too, because I would have never been so close to Him as well. God pulled me through. All the weekends that I came out here building this house, my memories were being healed because I was building it with the Carpenter from Nazareth. When I was framing or putting the shingles on the roof, He was healing me and talking to me about all my friends who died and that His plan for them was accomplished. He was there in the jungle with me, and He was here on the roof with me. I was a young man when I went to War; I lost my innocence in life when I had to kill people or be killed. I lost my friends. I saw the ugliness of war deaths. I saw evil face to face, but by the grace of God, I survived to be with your mother and to experience raising you and your brothers and sisters. Make your life count. Be a witness. Be the master of fear. Be a hero, raise heroes, and always protect your soul. Listen to the quiet voice inside your being. There you will find peace."

"Dad, do you remember when Elmer died? That sweet brown and white Cocker Spaniel was my best friend."

"Of course I do," Dad said.

"I was about eight years old and had gone to my friend Karen's house to play. We went down to the bayou where it had been dredged, and the mountains of mud had hardened and made these delightful mud castles. We had a great time running and climbing through those mud mounds. When I got home, I was in huge trouble for being late. I didn't have a watch, so I had no idea what time it was. Just as we finished dinner, Karen called. She said that Elmer had been hit by a car on Lake Street. I was stunned. You and I left in the car to get him. He lay with his head on the curb. I got out of the car and felt like I moved in slow motion. I stooped down and kissed his blood-matted ears. I was shocked by death. I wanted my best friend back. You lifted him up so gently and respectfully — I am astounded that after all the deaths you saw and aided in the War, you were so kind to Elmer. When we got home, you gave me the flashlight to hold while you dug a hole under the lemon tree. The whole family stood around watching the somber burial of our family pet. Mama said to me as we walked back to the house, 'Elmer went looking for you because you were late.' My heart was crushed like stepping on an acorn where all the potential life was gone forever. I didn't think I would ever be happy again. Mama and my sisters walked into the house. I turned the flashlight out so you couldn't see the tears streaming down my face because I was trying to be brave. You put your arm around my shoulder, and we quietly walked to the house. You knew that there were no words big enough for my grief."

Dad and I sat for a moment.

"Love killed him, Daddy. He came to rescue me. His love for me killed him."

Daddy nodded.

"It's the risk we take and the risk that soldiers take," he said, "Don't be too hard on your mother. She was worried about you. One of the things I like about her is that she is a straight shooter and will tell it like it is. I recognized back then that it wasn't the right time to reprimand you, and we talked about it later."

"Thank you for that kindness," I said, "but my point was that you were not the hardened killer that you thought you were. Daddy, you're a good man. You were good then, and you are good now." I whispered.

He choked up. I put my hand on his shoulder, and we both cried. To see tears in my father's eyes was huge because he was too much of a toughened Sergeant to cry. We sat a few moments in silence, emotionally exhausted.

-*-

"Dad, I have often wondered if I would be able to kill someone if I had been in your position. I don't think I could have done it."

"If you had been in my position and hesitated to pull the trigger, you would not be here today. We Americans value all human life, and we have soft spots for animals, which is a good and honorable thing, but you must know your job when you go onto the battlefield, and you must know the 'why' of your job. Let me ask you this," he leaned forward, "As a mother and protector of your children, if someone was trying to kill your child, and you had no other choice, could you pull the trigger?"

"Now you're bringing the mama bear out," I laughed. "But I have a confession to make. I'm bitter over the way the Marauders were handled by the upper military echelon. From the ego junkie who was General Stilwell to get back at the Limeys – the General's term – because of his first failed mission into Burma, the lack of planning, the lack of supplies, the lack of appreciation of the sacrifices of the soldiers and the disregard of the recommendations of the medical personnel to assess the health of the men. This would have saved many casualties. Do you have any bitterness toward the upper military echelon that made you expendable or to a nation that did not welcome you home with a ticker tape parade?"

Dad shook his head 'no' and leaned over and picked up his Louis L'Amour book and opened it. He took out a slip of paper that he had copied a quote from L'Amour.

Dad read, "Anger is a killing thing; it kills the man who angers, for each rage leaves him less than he had been before—it takes something from him."[46]

"I agree," I said. "But along with anger is forgiveness. This old saying says it all, 'Forgiveness is the fragrance that the violet sheds on the heel that has crushed it.'"

We sat silent for a few moments. I was letting that image sink in.

"Forgiveness is the true test of a warrior. Can you forgive the Japanese? It seems that humans have a long memory, and they rarely forget. Can you forgive the lack of planning and organization of Merrill's Marauders? Daddy, can you forgive me for being an insensitive daughter and not recognizing your PTSD?"

"You were not an insensitive daughter because I knew you loved me," Dad said. "You were not knowledgeable of what we went through, and I didn't tell you until now. Forgiveness and love seem juxtaposed. The more you love, the harder it is to forgive. Maybe that is the whole purpose of life—to learn to forgive. If you look at the world's unforgiveness, it feels a little hopeless," Dad mused.

"But by the grace of God, we can try!" I commented.

"Windy Lindy, I joined the military to defend our nation as our ancestors did. From the patriots of the Revolutionary War to the present, we gave our all. Ours was a secret mission, so not many people knew about it. When I attended the Merrill's Marauders Reunion, I became good friends with Lieutenant Colonel (then Lieutenant) Logan Weston, a man of God who could make sense out of the chaos of War. Weston, with his I & R platoon, was the sacrificial lamb that drew the Japanese up the Poakum trail to save the 2nd and 3rd Battalion. He should have received a Medal of Honor. Dr. Hopkins wrote under a picture of Logan, 'Lieutenant Logan Weston waits for D Day on the march to capture the airfield at Myitkyina. He has survived a trail-blocking operation against a reinforced battalion of Japanese infantry with a battery of four artillery guns. He and his I & R platoon, later reinforced by Lieutenant Smith and his infantry platoon, delayed the Japanese for four days. This heroic and brilliant blocking action allowed the

Second and Third Battalions to carry out their roadblock and get back into Nhpum Ga, where they could prevent the Japanese from outflanking Chinese troops. The action of Weston and his men saved the Marauders from almost total destruction. The Marauders would likely have been trapped, and the Chinese would have lost their spearhead and failed to advance further in 1944. Weston and Smith's platoons played a major part during the Battle of Nhpum Ga. His action and that of Lieutenant Smith, if under different command, could have justified recommendation for the Medal Honor.'[47]

"Logan told me that he, too, had an incident when a round hit his small Bible that he kept in his shirt pocket, and it did not penetrate. He also had a Japanese grenade land next to him, but it did not explode. I'm sure many more of the men, if they were alive today, would have come forward with miraculous incidents, as well as testimonials for awarding medals to the deserving Marauders. So many of the few records we had were either lost when a pack mule carrying them became a victim of a mudslide, or the records were simply not kept as General Merrill had requested," Dad said. He paused and then continued.

"Albert Einstein said, 'Great spirits have always encountered violent opposition from mediocre minds.'[48] The Merrill's Marauders knew we were fighting evil. We knew we were expendable, yet we fought on with courage, *esprit de corps*, and a kinship that bound us together with our very blood. We soaked the ground with our blood, but we defeated the Japanese in Burma. The Marauders were the great spirits, those sacrificial spirits for our country that built this nation. In August of 1944, the Marauders were aligned with the 475th Infantry. Later, in 1954, the 475th was retitled as the 75th Infantry. These brave and heroic men were the forerunners of today's Ranger Regiment. The 5307th Composite Unit Provisional code name Galahad was the unit of origin of the 75th Ranger Regiment of today. The Army Rangers follow the Marauder's tradition. We must never forget that as patriots, we must rise above the mediocre minds that often confound us, that threaten the essence of the United States Military and The United States of America."

Epilogue

A year after I interviewed Dad and he and I transcribed his diary, I remember getting a call from Mama that Dad was in the hospital. We drove to Lake Charles and visited him in his hospital room. My husband excused himself to call his office and left the room. I was alone with Dad.

He said to me, "You are bright and shiny, shinier than usual." Then he looked over my head and stared. I asked him who the nurse was standing at a counter that Dad and I could see through the doorway. He named the nurse. I knew he was lucid, but his eyes riveted back over my head.

I finally asked, "Daddy, what do you see?"

He began to name his friends and brothers who had passed away: Voris, Oscar, his Mama (mom Stephenson), and others that I couldn't understand. I think at that moment, he knew he was dying. I've only seen tears in my father's eyes three times in my lifetime. The first time was when we finished his diary, and we both cried. This was the second. Dad looked over at me with tears in his eyes.

That's when I looked at my father as another human being and not just through the eyes of his daughter. He appeared so young. It was one of the few times that I thought of him as a young man with his whole life ahead of him—his hopes and dreams. His hope of marrying a beautiful girl and having a family, which he eventually realized. He lost his youth—the happiness and fun of being young once. He was only seventeen when he joined the army. When all the college students were partying and going to football games, he was fighting in a war. When people got into spats and thought the world was coming to an end—he was being shot at, and his world was close to coming to an end. When people lamented about a lack of money for food, he ate K rations for four months and lived and fought through days of hunger before an airdrop could be delivered. After all he went through, I was amazed at how happy, resilient, and compassionate he was.

My brokenhearted family gathered around his hospital bed. We sang Dad's favorite song, "You Are My Sunshine," which my sister and I had sung to him many times when he was in a nursing home. My husband came in and baptized him before he slipped into unconsciousness. We assumed he had been baptized because he went to the Baptist church, but we were not sure. Later, we learned he had been baptized but failed to ever mention it to us. Our eleven-year-old younger daughter stood next to him with her child's wooden Rosary around her neck. She had a Kool-Aid-stained smile on her face and stains on her shirt. She stood next to her Papa, drawing hearts on his arm with her finger. My mother, his darling Evelyn, stood at the end of the bed. He looked at her, and then he closed his eyes for the last time. We lifted him up in prayer, and when he took his last breath, a huge tear rolled down his cheek. That was the third time I saw him cry, and it was the final time—the *lacrima mortise*—the tear of death. I considered it the cry of joy, the joy of being home and meeting the One he talked with all his life and prayed to, especially for safety and guidance in the jungles of Burma. There were no vultures waiting for his death. In my mind's eye, I saw Jesus reach into the foxhole, grabbing Daddy's hand to pull him out, and he was no longer bound to the ground and unable to escape. He was no longer tormented by the horrors of that War. I saw him being greeted by his family and many of his brother-in-arms. I saw him with his friend Tom Mattlock, who said, "I told you we would make it home, brother." My closed eyes did not stop the tears from running down my cheeks.

"Welcome home, Daddy—the war is over," I said under my breath.

-*-

My father passed away in 1993 from heart-related problems from having Orientia Tsutsugamushi Fever in Burma in 1944. He passed away at 73 years old, a young age compared to the longevity of his parents, who lived to be 99 and 100. Although I am grateful for the years he lived, I do wonder about the years he could have lived and enjoyed life because he certainly deserved it. After Dad passed away, I researched the subject of the use of penicillin in Burma in 1944. I found an article entitled "Scenes from Chinese – U.S. Offensive to Open Ledo Rd." in the "CBI Roundup" out of Delhi. It described "the treatment of a Chinese officer who was wounded at Myitkyina while fighting with the Chinese and American troops under General Stilwell." The following is a quote from the article. "... The wounded Chinese warrior is Lt. Jang Foo Chwan. He suffered wounds in both legs and one arm. Rushed by a hospital plane from the front, he was administered $2,000 worth of penicillin in a 24-hour battle by media [sic] against the deadly gas gangrene. Then, working in the surgery tent, ankle deep in mud, and with the monsoon rains beating furiously on the canvas, the soldier's right leg was amputated by Lt. Colonel C.J. Berne. Following the operation, Capt. Reuben D. Chier, assistant surgeon, administered blood plasma, a gift of the Blood Bank of Los Angeles."[49] I was elated that penicillin was used to save Chinese lives, but why couldn't it have been used for the Marauders as well? It took time for me to get over that disparity.

Time is the microscope for understanding our transformation. Time is also God's gift to us; how we use that time can be a gift to God, humanity, and the fulfillment of our destiny. However, timing can be understood in several ways: *Chronos* is the Greek word for time on this Earth that we generally think of as chronological events. *Kainos* is the Greek word for new, as in a new understanding. Dad was very good at entering chronological time in his diary. When he controlled his fear through prayer and other techniques, he responded to his challenges more efficiently, as if he were in the flow of not overthinking. The day he was hit in the chest, and the round did not penetrate, was it God's perfect timing in those circumstantial events? If he had dug his foxhole one step closer to the Japanese soldier, the bullet would have penetrated his chest. Or was the Japanese soldier in imperfect timing and location when he took the shot too soon instead of moving one step closer? If he had moved one step closer, the bullet would have penetrated. After this event, Dad knew that God was with him, and with God's help, cooperation, and Dad's faith, he knew he would survive.

I often pondered the story of the paralyzed man of Biblical times who was lowered through the roof so that Jesus could heal him. My first reaction was, 'Wow, they just tore up the roof!' but I came to realize that God wants the creative process and determination to meet Him from us, to prove that love is unstoppable. Love is in the realm of the eternal, and it is not limited by time because the essence of love is timeless and unstoppable. It is the only thing you take with you when it is your time to die. We will be judged by the measure of our love.

-*-

After years of petitioning members of Congress and the Senate by myself, members of the Merrill's Marauders Proud Descendants Organization, their parent organization Merrill's Marauders Association, attorneys, and fellow patriots, on September 22, 2020, Congress passed the Merrill's Marauders Congressional Gold Medal Act which was subsequently passed by the Senate. On October 17, 2020, it was signed into law by President Donald J. Trump. This award, which has been bestowed since the American Revolution, is our nation's ultimate expression of appreciation ratified by Congress and Senate for distinguished contributions at the risk of life over and beyond the call of duty. Robert E. Passanisi, the spokesman for the Merrill's Marauder's Association, said, "This recognition means so much to me and the other survivors and our families. My one regret is that only eight of us are alive to enjoy this historic honor."[50]

As the campaign ended, the 5307th Composite Unit (Provisional) was expended. There were so few men left that were able to stand that there was no last formation or acclamation by General Stilwell. But the men of Merrill's Marauders, along with their allies, the Chinese 22nd and 38th Divisions, the British, Gurgkas, and Kachins, knew that they had defeated the Japanese 18th and 53rd Divisions. The sick and bedraggled Marauders were triumphant on the airstrip at Myitkyina. The Mars Task Force was sent to join with about two hundred original Marauders who were sent back to fight to finish the task of ousting the Japanese from Burma, which was no easy task. My father and the remaining Marauders slipped back into American society with no acknowledgment or recognition. To them, they were just doing their job. They were expendable. The Marauders that I met at the

Merrill's Marauders reunions were very much like my Dad—extremely humble and grateful.

-*-

Although I wrote this book for my descendants, if you happen to choose this book and wonder what it was like in our day and discover that we are related by bloodline or simply as human beings, this book is for you. Much like you in your human condition, we are fearful of dying; however, we are magnanimous in suffering and sacrificing for a cause and for those we love. These are true events and people in a real war (World War II), which began on September 1, 1939, and ended on September 2, 1945. The Marauders participated in the action from February 1944 through August of 1944 in the harshest environment against a formidable Japanese Imperial Army and with shockingly scant food. The intensity of being under fire for that length of time could break the most recalcitrant soldier, but the Marauders powered on. The War took its toll on the soldiers who fought for their cause—soldiers who had talents, aspirations, and dreams of the future that were never realized. For their incredible mission, the 5307th Composite Unit (Provisional), popularly known as Merrill's Marauders, was awarded the United States of America's highest honor— the Congressional Gold Medal. The Marauders were the 172nd recipient of the Congressional Gold Medal since George Washington.

The torch is passed to you to define, enlighten, and preserve a peaceful future. Go forth, brave one.

Public Law 116-170— Oct. 17, 2020 134 Stat. 775

Public Law 116–170 116th Congress

An Act

To award a Congressional Gold Medal to the soldiers of the 5307th Composite Unit (Provisional), commonly known as ''Merrill's Marauders'', in recognition of their bravery and outstanding service in the jungles of Burma during World War II.

Be it enacted by *the Senate and House of Representatives of the United States of America in Congress assembled,*

SECTION 1. SHORT TITLE.

This Act may be cited as the ''Merrill's Marauders Congressional Gold Medal

Act''.

SEC. 2. FINDINGS.

Congress finds that—

in August 1943, President Franklin D. Roosevelt and other Allied leaders proposed the creation of a ground unit of the Armed Forces that would engage in a ''long-range penetration mission'' in Japanese-occupied Burma to— (A) cut off Japanese communications and supply lines; and

(B) capture the town of Myitkyina and the Myitkyina airstrip, both of which were held by the Japanese;

(1) President Roosevelt issued a call for volunteers for

"a dangerous and hazardous mission" and the call was answered by approximately 3,000 soldiers from the United States;

(2) the Army unit composed of the soldiers described in paragraph (2)—

(A) was officially designated as the "5307th Composite Unit (Provisional)" with the code name "Galahad"; and (B) later became known as "Merrill's Marauders" (referred to in this section as the "Marauders") in reference to its leader, Brigadier General Frank Merrill;

(3) in February 1944, the Marauders began their approximately 1,000-mile trek through the dense Burmese jungle with no artillery support, carrying their supplies on their backs or the pack saddles of mules;

(4) over the course of their 5-month trek to Myitkyina, the Marauders fought victoriously against larger Japanese forces through 5 major and 30 minor engagements;

(5) during their march to Myitkyina, the Marauders faced hunger and disease that were exacerbated by inadequate aerial resupply drops;

(6) malaria, typhus, and dysentery inflicted more casualties on the Marauders than the Japanese;

(7) Oct. 17, 2020

(8) [S. 743]

(9) Merrill's Marauders Congressional Gold Medal Act. 31 USC 5111 note

134 STAT. 776

Determination.

PUBLIC LAW 116–170—OCT. 17, 2020

(8) by August 1944, the Marauders had accomplished their mission, successfully disrupting Japanese supply and communication lines and taking the town of Myitkyina and the Myitkyina airstrip, the only all-weather airstrip in Northern Burma;

(9) after taking Myitkyina, only 130 Marauders out of the original 2,750 were fit for duty and all remaining Marauders still in action were evacuated to hospitals due to tropical diseases, exhaustion, and malnutrition;

(10) for their bravery and accomplishments, the Marauders were awarded the ''Distinguished Unit Citation'', later redesignated as the ''Presidential Unit Citation'', and a Bronze Star; and

(11) though the Marauders were operational for only a few months, the legacy of their bravery is honored by the Army through the modern day 75th Ranger Regiment, which traces its lineage directly to the 5307th Composite Unit.

SEC. 3. CONGRESSIONAL GOLD MEDAL.

(a) AWARD AUTHORIZED.—The Speaker of the House of Representatives and the President pro tempore of the Senate shall make appropriate arrangements for the award, on behalf of Congress, of a single gold medal of appropriate design to the soldiers of the 5307th Composite Unit (Provisional) (referred to in this section as ''Merrill's Marauders''), in recognition of their bravery and outstanding service in the jungles of Burma during World War II.

(b) DESIGN AND STRIKING.—For the purposes of the award referred to in subsection (a), the Secretary of the Treasury (referred to in this Act as the ''Secretary'') shall strike a gold medal with suitable emblems, devices, and inscriptions, to be determined by the Secretary.

(c) SMITHSONIAN INSTITUTION.—

(1) IN GENERAL.—Following the award of the gold medal referred to in subsection (a) in honor of Merrill's Marauders, the gold medal shall be given to the Smithsonian Institution, where it shall be displayed as appropriate and made available for research.

(2) SENSE OF CONGRESS.—It is the sense of Congress that the Smithsonian Institution should make the gold medal

received under paragraph (1) available for display elsewhere, particularly at other locations and events associated with Merrill's Marauders.

SEC. 4. DUPLICATE MEDALS.

Under such regulations as the Secretary may prescribe, the Secretary may strike and sell duplicates in bronze of the gold medal struck under section 3, at a price sufficient to cover the costs of the medals, including labor, materials, dies, use of machinery, and overhead expenses.

wwoods2 on LAPJF8D0R2PROD with PUBLAW PUBLIC LAW 116–170—OCT. 17, 2020 134 STAT. 777

SEC. 5. STATUS OF MEDALS.

Medals struck pursuant to this Act are national medals for purposes of chapter 51 of title 31, United States Code.

Approved October 17, 2020.[51]

Merrill's Marauders Medal, from the United States Mint

The following are postcards from Chinese school students who studied the 1944 CBI campaign in their history classes. They wrote to the surviving Marauders as well as the family members of the deceased thanking the Marauders for their contributions in defeating the Japanese. The following are the postcards that my family received.

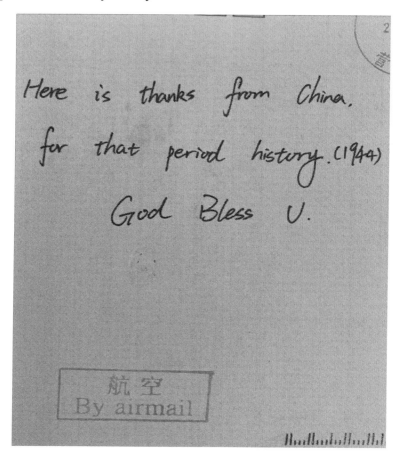

The student writes: *"Here is thanks from China for that period history. (1944) God Bless U."*

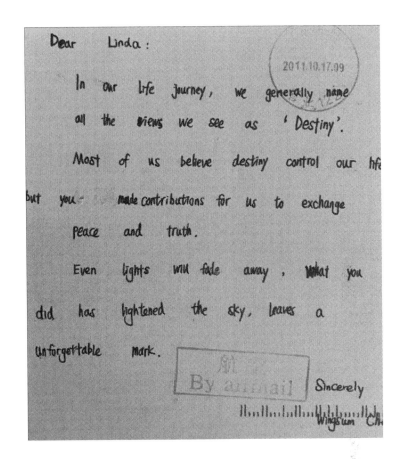

The second student writes:

> "Dear Linda,
> In our life journey, we generally name all the views we see as 'Destiny'. Most of us believe destiny
> control our life but *you made contributions for us to exchange peace
> and truth. Even* lights *will fade away. What you did has brightened*
> the sky, leaves a unforgettable mark. Sincerely,
>
> Wingsum Chen"

I was touched by the lyricism and sincerity of the sentiments of these students. Yes, Merrill's Marauders made a brighter future and left an indelible mark on the history of WWII by stopping the Japanese aggressors in the China, Burma, and India Theatre of War.

Front of postcard

The following is the complete 1944 Diary by Larry W. Stephenson. I did not make any clarification, grammar, or spelling edits but typed it as it was written, even paying attention to the upper or lower case of each letter. There were numerous areas that were smudged or had a water stain which made the words indecipherable. In such instances, I simply put a line.

Linda S. Cunningham

Diary of Larry W. Stephenson

5307th Composite Unit (Provisional) Merrill's Marauders

Jan.1. Regiment is formed. Parade formation 10AM. Usual camp duties. ___ ceremony per per 1 50 GQ See Above

Jan. 2. Usual camp duties. Police of area. Col. Hunter inspected area.

Jan. 3. APO884 c/o/ PM NYNY Left camp Deograh 0730. Marched 9 mi. Arrived 1800 at Baroda bivouac AR Tactical formation. 3rd plt. protected P & d in blowing tracks. 2nd and 4th went with column. Attacked Jocklaun 4:03 Left Jocklaun 16:15 for ___

Jan.4 ___ Action

Jan. 5. Aironia, India. Left Baroda Bivouac Area 0435. Arrived at Aironia 0730 Distance marched 1 Tactical formation. Attacked airfield. Aironia 1st and 2nd platoon held up by heavy fire. 3rd and 4th platoon was successful. Wasn't as good as to be expected. Bivouaced at Aironai. Sent put Crump from DY to camp Deogra. 4th plt Ration drop1640. Left Airaonai. Arrived 1815 Attacked Hana 2 miles Tactical Form.

Jan. 6. Beora, India.Left Hana Hana 0630 arrived

Basra 0745. Distance marched 2 mi. Attack on column. Assembled & moved out. Column formation. Wash up and rest. Plt 1 4 position, Disperse. Had plt filtered thru lines. 0800. Left Basra 2000. Attack on B Column. Bivouac area. Nibaho 0700 arrived. Miles Marched 1 mile.

Jan. 7. Cp. Deogarh, India Left. Nibaho India 1700 arrived. Miles marched 12. Notes left Basra 0200. Infiltrated thru enemy lines. Assembled 1000 yds from Surauwa. 225° Attacked B Column 0725. Successful. Reassembled. Had chow. Rest until 1700. Co. HQ ½ captured. Cooks and 1st Sgt got thru arrived. 0200.

Jan. 8 Cp Deogarh Rest and usual camp duties. Rained

Jan. 9 Cp. Deogarh. Usual camp duties. On range M I. carbine BAR

Jan. 10 Cp Deogarh. Usual camp duties. Fired on range.

Jan.11 Cp Deogarh India. Usual cp duties. Plts fired on range. Plt & Sqd problems.

Jan. 12 Cp Deogarh India. Usual cp duties. Plts fired on. Plt and sqd problems. 1st plt puton problems. Lord Mountbatten gave lecture.

Jan. 13 Cp Deogarh India. Usual camp duties. Plts fire on range. 4 reductions.

Jan 14 Cp Deogarh India. Usual camp duties. S.M.G training. Mortar & snipers. Packboards & pins turned *in.*

Jan 15 Camp Deogarh India. Warning Alert. Usual cp duties. Was rescended. Had platoons fire.

Jan. 16 Cp Deogarh India. Usual cp duties. Turned in Barracks Bags and drew D Bags.

Jan. 17 Cp Deogarh India usual cp duties. Non swimmers practice swimming.

Jan. 18 Cp Deogarh India. River crossing by Bn.

Jan 19 Cp Deogarh India. Usual cp duties Platoon tactics.

Jan 20 Cp Deogarh India. Usual cp duties. Trfd to L

Jan.21 Cp Deogarh. Usual camp duties. Good outfit.

Jan. 22 Cp Deogarh _____

Jan. 23 Cp Deogarh India. Usual cp duties. My reduction came in today. Not much doing.

Jan. 24 Cp Deogarh India 1st Bn pulling out tomorrow.

Jan. 25 Cp Deogarh India. Fired on range. Lots of fun.

Jan. 26 Cp Deogarh India. Usual camp duties. Cleaned up area. Went to River & washed clothes. Letter from Charlie. Wrote to Evelyn

Jan. 27 Cp Deogarh, India. Fired again today. 3 natives shot by Gunn and Nichols 1 died. No letters. Saw show Clark Gable in Will Meet You Somewhere and also

A Yank at Eton and Holliday Inn.

Jan. 28 Enroute to New Station. Cleaned up camp. Assembled near camp. Cleaned up camp. Poor meals. Started to rain. Left out 745 Arrived 1120 WesterHauson had car all ready. By Plt 3rd plt has swiped 10-1 rat. Sick as hell. Left out at 3AM.

Jan. 29 Enroute on train to New station. Left Jochlin India 3AM. Had early breakfast. 9AM Then cleaned rifles.

Jan 30 Enroute on train to New station. Stop and Go. This damn train. Passed thru hundreds of towns going S. E. & S.

Jan 31 Enroute on train. Still stopping and going. Lots of small towns will arrive in Calcutta ___t nite.

Feb.1 Enroute to new station on train. Arrived Calcutta changed tracks. To narrow gage. 5 quarts whisky.

Feb. 2 Enroute to New Station. Off of train at 0700. New camp. Rest and wash up. Saw ferry tomorrow.

Feb. 3 Enroute to New Station by ferry. Up 0600 moved to Ferry 0900 for 36 hour trip up River. Brahmaputra. Slept colder than hell on stern of Deck.

Feb. 4 on Ferry enroute to New station up river all Day. Nice & warm. Some mts & Fishing. Tempas Had picture taken 4 times. One goes to states.

Feb. 5. Enroute to New station by Ferry. Slept starboard mid ship. Under Johnston. Pretty warm 10-1 ratns. Off of ferry at 10AM Packs to camp in Assam. Then we loaded. Hold lot of weight. Nice place.

Feb. 6 Rest camp in Assam. Just across River Chinese moved out at 0800 packed up and Ready to go on train. Bombing 200 miles fr here. We are going. Boarded train at 1200. 10-1 ratns. Cold soda. In same car slept on air mattress on floor.

Feb. 7 Enroute to New station on train. Up 700 Cold as hell. Had good breakfast. Still traveling toward Tsikudio East. Was given box of tangerines by drunk GI. Still more jungle on train.

Feb. 8 Enroute to New station. On train to Assam. Rode all day. Ate & ate 10-1 Rations. Slept colder than hell. Saw Chinese girls. Good shanks. Off of train at 5pm. The piled into 3rd platoon Bld. Slept nice.

Feb.9 Sat Enroute to New sta. Up 7AM. Had Reveille. Took much needed shower. Moved out 1130. Doeling & I were late. Put into barracks in jungle in Sgt. Doeling's tent.

Feb. 10 Sund in rest camp. Had to take shots. Typhus. Then was made sgt in 4th plt Mulligan and Lt. Burk. Rolled pack. Very heavy. Moved over _____

Feb. 11. Mond Enroute to New station. Left bivouac area with 4th plt. My squad is point for 5th dispersal group. Moved out at 820 pm. Tired & Sore. Rained for 19 miles. Sleep at 430. Up at 930.

Feb 12 Up 930. In Bivouac area on Ledo road. 7 miles fr Town of Ledo washed up. Rested cleaned MI.

Rained. Started to move out again at 645 marched to 2nd bivouac in shacks. Arrived 130am. Very Tired and Sore. Rained. Uphill and downhill. Rough 16 miles.

Feb. 13. Enroute to New sta. Arrived 130AM. Miles marched 16. Left Bivouac ARea 530. Feel better. ARRived 3rd bivouac area 1145. Feel a little better. Slept in barracks next to Lt. Bert. Air mattress is swell. 2 meals per day rough. Miles marched 10. Left camp at 5:30. 2nd Dispersal group under Sgt. Mulligan. Arrived 4th bivouac ARrea at 830. Miles marched 2.

Feb. 14. Enroute to new sta. Saw Melvin Douglas in show. Had grapefruit juice last night. Slept cold. Still on Ledo Rd, lots of traffic. Cut my squad's hair today. Very poor Chow Rested. Tonight we leave cp at 530 & will be in Burma Road at 1130. March wasn't too rough. Covered 12 miles. Feet are not 2 bad shape. Arrive in Tanchu Rest camp. Had movie & good coffee. Slept well. Rained like hell. Men got wet over the Big pass. Cold as Hell

Feb 15 Enroute to New sta. It was rough Last nite & cold. Left Bivouac on mt at 530. My squad is on Rear of 2nd dispersal grp. Miles marched 12. Saw movie. Coffee & Tea Bed at Tincho 59.25 Miles fr Ledo. Washed up During day & wet shoes and towel. Read book & then left out again at 630 in 4th dispersal group. Enroute to 7th bivouac. Am wearing jungle boots tonight. Try them out. They didn't work so well. Arrived at bivouac area at 1130. Tired as hell. Distance marched 12 miles. Bed

1pm.

Feb 16. Up at 8am. Had poor breakfast. Cold last night. Slept 3 hours then cleaned up. Have bad knee. Left cp at 600pm. New cooks and mess Sgt. Moved out at 820. Arrived New bivouac area at 945. Bed at 10. Crossed river _____

Feb. 17. Up early read book on Action at Aquilla. Washed up. Good Breakfast. Will move out early. Miles ____ 4th dispersal group. The hill was really rough. Had tea at negro camp.

Feb. 18 Up 8AM on top of hills. Clouds below us in valleys 2 more miles uphill. Had fair meals. Everything tactical from now on. Moving out at 315 for 7 miles. Left out and marched 8 miles instead. Boy, am tired. Arrived at 930. Bed and sleep. Up hills pretty rough. 6 miles downhill.

Feb. 19. Enroute to New ARea. Up early. Rested and ate breakfast. Moved out 445 4th Dispersal group. Arrived 1030. Mahmood and rest had 7am guard. Ate poor supper at bridge Bivouacked on top of hill. Gorretti slept in Negroes tent and drank beer.

Feb. 20 8 miles. Up early Down to breakfast at curve in road. Goddard fixed me up to sleep in native shack. Swell Breakfast and tea in. Rested well. Mahmood, Goretti, Roger & Bert say should be in tonight, just about to FootHills. Hurray. Moved out again at 11am. We are now the 5th dispersal. Went 11 miles and boy was it rough. Mud Rain and everything. Bivouacked beside road between big logs. Rested and it rained like hell. Arrived at 3 pm. Got coffee and extra chow at kitchen across bridge by Chinese camp. Rained.

Feb. 21 Up 7 am. Everything is wet & we are cold Air mattress was swell but is Heavy Goddard gave me musette bag Have a very heavy pack. Left bivouac area 10am. Marched all Day Boy its Tiresome. I'm sore. Shot one mule. It's wore out. Just 5 miles from Japs. Marched 18 miles. Arrived 9pm at Paddy field. Dead Japs all around us. Chinese too. Colder than hell and tired.

Feb. 22. Up 7AM Rolled packs & Bivouacked 600 yards in Jung cleared out place for our squad. 1st one ___ ___ & Bath and washed clothes. ___ Bn is on 2nd watch In River Dead Chinese and Japs all over the place Given 10-1 rats. Found wild Lemons Have swell place for camp. Smuck is good man. Gurkha knife comes in handy. Bed Early.

Feb. 23. At bivouac area near Nainka, Burma at edge of ____ River. Slept, cleaned up. Put shelter up we are moving out tomorrow Turned in shortage of squad. Had 4 cups of good old coffee. Expect bombing raid.

Feb. 24 On our mission up at 630. Cold as hell. Started to move out at 9AM. Hold up until 11AM. Passed river where 900 Chinese were killed. Thru 12 Jap positions and marched 11 mile thru jungle. Was very thick. Arrived in bivouac area Beside Trail at 5:30. Coffee and meat and beans. Not bad. My squad had right flank.

Feb. 25. Up early 630. Jap planes flew all night. Had stew and coffee and coffee for Breakfast. Moved out at 9

AM. 4th dispersal group.1st Battalion c company. All the old boys. I & R contacted Japs. One of our men shot thru nose. 1130AM 1 Jap killed. 8 miles bivouacked. Heard lots of mortar and BAR fire.

Feb. 26. Up at 630. Moved out at 730. Nothing doing last night. Just mortar fire & BAR. Was held up for 40 min By 20 japs on patrol. ARRived at New ARea. Marched 11 miles. Set up defense on left of Bn. 1man KIA. Fixed chow. Short of it guards say Jap patrols active.

Feb. 27. Up and at em at 630. Move out at __ I Led 4th dispersal group at Tanga ga. Our ___ over at 1230 Marched 6 miles Up and then ___Straping area ___Cut my finger bad. Ate supper very Hungry. Rtns tomorrow I Hope Would like to hear from the little woman. On watch at night. Fired BAR at nips heard them in Brush. Slept well.

Feb 28. Up 9AM. Cooked very frugal meal. Bouillon and crackers and plenty of coffee. Dried out pack and equipment. Had air drop. Rations. Ate in a hurry then started out for. Waded 5 rivers and 54 creeks. Left at 2 pm. Bivouacked at swamp. Colder than hell. Too much noise and lights. Expect patrol activity. Slept cold as hell.

Feb. 29. Up 6AM Built fires. We protected Bn Col Beach left out in 4th dispersal 500 to 800 Japs on our right flank. Enveloping move. Left out at 815. Crossed Chinese at least 9 more times. Sore as hell. Tired and wet. Marched 16 damn miles. Arrived in bivouac area at

8 PM Cooked supper. After dark dangerous. Nip planes flew over. Bed 12PM. Tired.

   Mar. 1. Up 630. Cleaned up and rolled pack. Men had to shave. 4th platoon on detail for air dropping. Cooked only Bouillon and a crackers and hot pepper. Feet wet will dry before moving out. Surprise no dropping. Built up stacks of wood for signal fires. But had to leave out at 730pm Marched all night thru black jungle trails. Tired and sore as hell. Miss my woman. __ girl

   Mar. 2. After marching ___ miles & crossing 5 more rivers. 0820 12 hours of straight march. Kinda rough. Ate breakfast. The mg & mortars Raising hell. Just a Short Distance Away. Slept for 1 hour in 48 hours. Awake and on move. Moved out after supply drop at 5pm Thru 6 miles of Jap positions. Bedded down and Lieut. moved us to another place. The jerk.

   Mar. 3 Up 545. Got pack. I & R 1pm Go into motion moved to trail block for 20 min. Japs hit us. 2 men were wounded by shrapnel. Attacked again at 620 this morn. Moved out as Rear point for column at 820 got within a mile and ½ of our objective and we hit the Nips. 7 killed. I don't know now many was wounded. moved on and bivouacked. Sleepless night.

   Mar. 4 Up at 600 Lots of firing all night. Started to move out as point. Fighting all around us Only God can bring us through. Moved out to relieve I & R. Snipers just about got our squad. We got __ Jap I was the point. Dug in across river from Japs. Had a hell of a fight. 2

men got killed. Had ration drop but no food all day. Drank 3 cups of coffee. God saved us. Thank the Lord. Always trust and he will help. Expect heavy Jap fire tonight. B GH coming up road s with tanks A BN in reserve. Chinese on the way. God, I hope mail comes in. I miss my darling. Would love to see her again. Please God send me back to her. She needs me, Please God.

Mar. 5. Still across river from Nips. Small battle last nite. Col. Beach led patrol across and just made it before all hell cut loose. Threw hand grenades. Still no rations. Fixed pack. Not much sleep. Dirty. God take care of us. No chow for 3 days but still going. Dive bombers opened up on Nips with bombs and shell fire. My squad and I went on patrol into Jap lines 4 miles. 1 man hit by sniper. Ver lucky. God was with us. Slept well.

Mar. 6 Up 620. Shelling by mortar and mg all nite. Getting used to it. 4th plt Moved out. 9AM to protect Air Drop. Back again at 12 noon. Jap mortars were shelling position very close. Finally got a Little Rations. Not bad. Mahmood and Gorretti went to Front. Dive Bombers Strafed and Bombed Enemy positions. Volunteers to get water. 530 Nips attacked for 1 hour. It was hell. They ran at us by waves But our mg cut them down by the hundreds. Bullets cut the ground & leaves all ARound us. Artillery mortar, machine guns & snipers just about got us out of ammo. Layed awake until 330 sneaked out of positions and withdrew. 3 Letters to mail. 3 hours sleep. Only God could have got us thru.

Mar. 7. Hit the deck 7AM. Mortar and mg fire still on. Moved up to trail where air drop was then drew rats and moved out at 12 noon. Man was Killed at Air Drop & buried. Moved 3 miles. Arrived 5 pm. Cooked up chow. Found Rice and corned Beef. Tasted swell. Bed 8 pm. Saw old buddies and Chambers good boys. God pulled us through.

Mar. 8. Up 7AM. Nice and Sunny. Drew Ratns went over and saw the pictures that Ethel sent. Were swell. Food Ham & fish Rice & Bouillion 1st Bath after 2 weeks. Swell Rested pulled out at 12 Noon. Marched 7 miles. Arrived at Bivouac ARea at 820. Bedded down. Contacted 10 man patrol.

Mar 9 Hit the deck 7AM moved 300 yards to River Trail. Set up defensive positions. Ate 4 times Rice and other things. Sick as hell. Took bath and cleaned up. Have Dysentery. Sick as Hell. Lt Wyngardner is new plt Leader. Lt. Burk was reld.

Mar. 10. Up early 620. Everything alright so far. Sicker than hell. Stomach weak. Also ate chow twice. Was issued ratns for two days, Heard news broadcast publicized Yanks in Burma. Had Air Drop. General & troopers went to Mangwa. Bombed for year and ½. Not much else.

Mar. 11. Up this morn weak as Hell. Chest and kidneys hurt. Got Air Drop 10-1 rations. Ate quiet a bit. layed around. Took bath and washed clothes. Got clothing issue. We are to be on the left flank moving out tomorrow at 10AM. Still feel like hell. God please give me strength to go on. Slept under parachute.

Mar. 12. Hit the deck 6AM. Policed up area. Dried

out clothes. Fixed pack. 4 days rations. Big load. A Bn came by at 9AM. Saw all the old goys. Good to see Them. Still weak as Hell. Hope I can make it. Moved out at 1030. Behind 3rd platoon. Marched 16 miles. Rough as Hell. Crossed river and arrived at LaJauge. 545pm from Ninglap Ga. At breakfast food, coffee and went to bed on air mattress. Chinese fired in to bush and killed cows. Feel better.

Mar. 13. Hit the deck 63 AM Dried out and made pack. Ate soybean cereal will move out at 9AM Have 6 days rat because I couldn't eat. The Burma Raiders will march 12 miles Today To Kadau ga. cross 5 Rivers. Followed to Padum. Boy it was Rough. In at 445. Marched 31 miles. Rough as Hell. Tired and sore. Passed Last Chinese out post. Took cold bath in Tana ga. Ate supper, to bed at 8pm.

March 14. Hit the deck 630 Still have 4 days rat. very Heavy. Packed up. Slept swell. Ready to move out. Moved out at 945 went 8 miles. Kinda slow and tiresome. Into bivouac area at 3 pm. Built fire and dried out clothes. Bed early. Very sore.

March 15 Hit the deck Early. Men were issued shoes and rations for 3 days. Moved out as point. B Bn is in Front of us. Khaki in rear. Marched 7 miles. A lot up Hill. Rough. Nice Bivouac ARea.

March 16. Up and at em 730. Left out as rear guard. Passed B Bn 6 miles up. Regt HQ. Had very bad Hills. ___ Arrived at Bivouac Area 0820. Marched 16 Roughest Hills & miles I Have Ever Been on. Rained like

Hell. Got wet. Bivouacked after Eating. Bed late. Had guard.

March 17 Up 5AM Packed and left out AM. Marched 4 miles for air drop. Johnny Kazoski was killed. 5 holes in him. Saw doc work on him for 2 hours. Dried out clothes and blanket. Took Bath. Was issued rations K & C & candy. Swell. Ate good chow. Had a Hell of a Stomach Ache. Built nice Shelter with Banana Leaves. Smuck and I. Rained like Hell. Kept Dry.

Mar. 18 Up early 630. Fixed pack. Found out we are staying Here. Ate meat and Bean stew. Darn good. God is Kind. Layed around. Not much doing.

Wish god would send some Mail. Have been getting a good Appetite. Bed 830. We will be out of Here in 20 days. Got parachutes to sleep on.

Mar. 19 Up 7AM Had nice sleep. 4th platoon is on airdrop detail. Drew 3 days rations for 2 days. Expect mail today if God is kind. Ate c rations, got water shaved and aired out my feet. Miss you honey. Had trouble with McBride. Went to river with old dad Goddard. Washed shorts and socks. Took bath. News we have completed one more mission. A Battalion lost few men. We have one tough battle ahead. Ate stew, cheese and bouillon. Rice and coffee. Feel fair. Move out in morning. Early,

Mar. 20. up at 715 rolled big pack. 4 days rations and ready to move out at 8AM. Finally moved out 11AM. Marched 3 miles all up hill. Roughest yet. Just barely made it. Arrived at Janzan are on Roadblock trail leading NW. Fixed shack. Have 2 BAR's on trail. 3 hour guard 12-3 AM. Bed early.

Mar. 20 [21] Up and at EM at 0600 Woke up Mahmood. Last shift. Cooked crackers, bouillon, pork and egg yolk & cheese. Tasted swell washed Down w coffee. 2nd Squad takes over Trail Block 12 noon. Were at Janzan. 25 miles from Shadazup. A Bn is There. 2nd Bn & Khaki column is at Warong. 25 miles. We will get Airdrop of 10-1 Rtns Today. 3 pm got chute and 3 Extra Rations Then 1 Box & Half 10-1. Ate till I could not Eat more. Fixed pack and Bed. Couldn't sleep.

Mar. 21[22] Up 545. Made pack with 4 days of c RTNS & Extra candy &stuff. Ate good old 10-1 for breakfast. Pulled out at 8 behind 3rd pltn & Bn Then us. Have 9 miles to go. Darn rough marching up Big Hills and Down. Pulled in at Bivouac area at 3pm. And put up trail bloc leading to Warong and Kaming. Rained like Hell. Gosh got letter and more letters from Evelyn 24 letters. Fr 1- 4 letters and Paul 1 and Mr. and Mrs. Hower 1 letter. 32 in all. Good news from CBI on Merrill's Marauders. Big Deals.

March 23 Hit the deck and moved out towards Warong at 7 AM. Rough as Hell up Hill and Down. Had to detour A Round Warong. Turned off Trail to Right covered 14 miles. Bivouacked After crossing 8 rivers. Nice fire, food and sleep. Native guides. Slept near us.

Mar. 24, Again Hit the deck and started out at 730 as Rear guard. We went 9 miles and put up Rear Trail Block near small Village. Expect Air Drop Tomorrow. Waded 16 Rivers. Wet and Tired pulled iN At 1130. Built fires and Ate good-old c Rtns stew. Don't know what is Next. But God will guide & protect us. in valley now near Kaming. 4th plt moved out to packum 5 miles. Stayed with I & R. Nice. Rest. 1 15 minute guard.

Mar. 25. Hit the deck at 600. Ate breakfast. Have 8 rough miles to go. All up Hill. Its going to be Rough as the Devil. Marched 8 miles to Warong. Have trail block to Ashau. The I & R was hit 2 times. Killed 9 or more Japs. No losses. We are in a Bad Spot. Mahmood went Ashau to Take Message. They sent Rtns by mule to us. Just About out. God will Take care and protect us. Ate mostly Bouillion and coffee. Bed 7pm Fresh Jap tracks.

Mar. 26 Up 5-7 on guard. Kinda Rough. Moved up Hill To Another Trail Block To Ashau & Dug iN. Thank you God for the protection & guidance Last Night. Ate Breakfast Then Nips Hit us with about 600 men. Boy it was Rough. Total Nips Killed About 25-28. We pulled out. Fast withDrawal. Boy it was Rough. Haven't washed in days & am Tired and Sore. Will Make Last sSand At Auchie. 3 miles from Warong. I pray to God to give us protection & carry us safely thru this campaign. God in Heaven give us Help. Please God send us Food and Ammo. B Bn came in at 1030. C Bn came in At 3pm. Have A Very good perimeter. Got 5 days of Rtns By Air Drop. No sleep Again Tonight. Nips all Around us. Boy its Rough as Hell.

Mar. 27 Hit the deck 530. Rolled packs and started out at 6AM. I & R, 4th pltn, and Rest. Boy we Ran 9 miles. Artillery, mortar, mg & Everything. I Ran till I was out then they Tied me on a horse and my pack got

Lost. Arrived in Sam Shing Yang 3pm. The Doc worked me up. Out. Only God can see me through. Please God in Heaven help me get safely thru this campaign and safely back To Evelyn's sake. McBride & I slept in fox Hole. No Quilts or nothing. Cold as the Devil. Swagger and Smith helped me. Thank you, Lord.

Mar. 28 Up 545. Cold as the devil all night. Drank coffee for Breakfast. Couldn't much more. Very weak. Helped in Air drop then 1st squad went up River and formed Ambush. Gorretti Brought chicken and pie up. Swell. Moved back to positions at 530. Fell asleep in parachutes. Mac and I both lost packs to Nips. Drew 10 and 1 Rats and clothes.

Mar. 29 Up 530. 4th plt is going up trail to contact B Bat Got Trapped & stayed until 2 pm & Then Had to Run Down to River. Just made it By 5pm. God was with us To pull us out ofAa Tight Like That. Made smaller perimeter. 3rd plt moved up to secure Trail. Rained like the Devil.

Mar. 30. Moved up trail to Kauri to get B Bat out. started raining. mg 31 Hit us 3 times. Just missed Mahmood & Taylor & Paul & I. Moved around Flank 3 Hours. Cole got Hit & 4 other guys. Rough as Hell. God please guide and protect us and get us back to the States safely for Evelyn's sake in Jesus Name Amen. formed perimeter on Trail at 530. Cold, wet. Raining. God was the only one That Saved us. Didn't Sleep, Too cold.

Mar 31. Up 530. Started up Trail At 6. Ran into mg 31 Again. Graham got it Again in ARm. Jack Pedro was killed. God please guide & protect us in Jesus Name. Amen. Big Robinson Robby was killed. 7 men were wounded and 1 killed By Knee mortar. All were Radio operators. Can't go forward. 2 75's were dpd To Be used by us. P-40 straffed and Bombed positions But it didn'tHhelp. So far the trail is clear. Cold as Hell.

April 1 This may be April Fool's Day But there is No fooling ARound Today. Abn patrolled Right Flank contacted Nips. There all Dug iN. Khaki came Thru with Bed Rolls. No chow as yet. Still cold. Had Set Ambush on waterpoint. Was relleved. God in Heaven pleas guide & protect us & bring us safely Through This and Take me back safely To Evelyn for Her sake. In Jesus Name Amen. Got Blanket & poncho. Rice was sent up by Khaki. Tasted swell.

April 2. Up 5AM. Expect Nip attack. Cold as the Devil. Cooked cup of tea. Swell. Some more Rice from Khaki. Still Trapped. God Brought us Through The Night safely. Expect they Break out Today w/ Gods Help. On water Ambush. Will Attack at 230. 4th went ARound lLft Flank for Diversion. Expect Dive Bombers & ARTillery BLast Loose At Them. We Fired & got the Hell out. 1st squad Led. The Nips followed & Fired on us. Bivouacked on Hill. Cold as Hell. God is w/ us.

April 5. Some How, I missed 2 days. The Lt led us in a circle Trying To get out. Then Barney Led us on The Trail. MaJor Lou was wounded. Capt Burch got Jap m.g. Tired & sore. Artillery & planes Raising Hell. God be w/ us in the coming Fights. saw Mosier, swell guy. got me water. went up & Built up perimeter Around 75's & 81 *mortars Then Built up perimeter At Foot of Hill Next to Nips. Didn't sleep very much.*

April 6. Up 5AM Rolled pack. Big push Today. Boy I mean push. The Artillery & Mortars & Dive bombers Laid Down A Big Barrage that very close to us. Deafening then All was silent, & we moved ARound Right Flank to protect Lt. Wyngarners plt. moving up trail thru Jap positions. Lots of Dead Laying ARound where our Dive Bombers got Them. Saw them Rush thru woods. We protected Trail to Rear. Gained ¾ mile. 1st squad went up to Feel out Japs on Hill. They cut Loose with mg & Rifles. It was Rough getting out of There. Bullets All Around us. Tayler got burned. 1day rat. Dug in. Mc & I had Tea, cheese & pork & eggs. Our ART shelled Nips.

April 7. Up at daybreak. Khaki is pushing on thru us. Orange is to protect trail. 4th plt is to follow Khaki and secure Trail. Please God guide and protect us in our coming trials. Thank you for keeping us safe up To Now God. The 1st sgt was killed. The other sgt wounded last night. 4th was commented on. Fab Job we Have Done and 1st squad as point. Dive Bombers, ART & Everything Hit 3 Times and we Attacked 3 times. We finally took First part of Hill. Moved Back to perimeter. Had tea, Bouillon & cheese. Feel swell. Dirty and contented. Had cgts and rats issued. Expect Nips to move out. Bed early. Slept well. 1 Hour of guard.

April 8. Up & at em at 5AM. Colder than Hell. A Bat is going on to right. Khaki up the trail & we are going Around Left Flank. Started at 8AM. Boy its Rough going thru Jungle up Hill and Down. finally Bivouacked on nose of Hill overlooking Town. B BN is in. Have Perimeter. God please guide and protect us. Carry us safely thru this campaign and Back To the States for Evelyn's sake. In Jesus Name, Amen.

April 9. Up 530 AM moved S. & west to Hit Nip Flank. Jungle is Rough *going* All Day Long. Reached our place at 2PM, Found out that Khaki had moved up within 75 yards from B Bat. Moved Back out Again. Up Hill and Down and contacted Nip patrol. Moved back to Trail & Took up old position to protect Trail. B Bn got thru and took out 147 wounded and 36 Killed. Rough as the Devil. Spread out thin.

April 10. Up early. Sgt Mason Has Strong point. Went out on patrol S. & West. Went out 2 hours. Saw Jap Tracks at water point. Rough as the Devil. Found Lt. Smaules Bivouac ARea for last Night & His Tracks Heading South went up to Nabum ga & got extra rat for men & Saw Dead Horses and Japs all over. ARea is shelled all over. Mulligan & Smith think they are Dictators. Not worth a Shit. Tonight, we are on A strong point. Just 6 men 250 yards on each side. *Rough.* God please protect & guide us safely thru This. Its going to be Rough going for Evelyns sake in Jesus Name. Amen

April 11. Up 5AM. Eyes and Ears open. This outpost is dangerous. Ate good Breakfast. Egg *yolk*, crackers & coffee. I pray to God to be relieved Before the Day is over and please God protect and guide us safely Thru This. In Jesus Name, Amen. Plt of Nips approaching us from the East. they were bombed and straffed. B Bn pulled out & B.C.T & A BN was put in. Everything looks black. 2 hours on outpost 12-2. Ate cheese & lemon and lots of chow. Need Bath. 2 weeks since I had one. Lots of noise. Think Nips occupied positions 75 ft Across Trail. Not any sleep.

April 12. Up 5AM Rolled packs. No Nips, I guess. Got Radio fr Bn call on Hour patrol Every 2 hours. washed Face Teeth & hands. Ate cheese & coffee. Outpost w/ Smuck. 10-12. Mason was busted. Good boy *screwed* By Mulligan & Smith. No good people. Pulled in outpost 4pm. Everything okay. Go on patrol up Trail. We are going to be relieved The 15th. Its wonderful or at least, I hope. Ate Bouillon, cheese & coffee. Bed at 730. Guard 8-10. 3-4 gets Kinda Rough. No sleep. Mahmood went to B Bat Hill & got medicn for leg sores. Sent up 1 days rat. No water as yet.

April 13. My Bad Luck Day. But I'm praying to God to see me safely Thru This. Goddard and McBride went up to Hill. It looks like Rain. Rained Last nite. Kinda Rough. Taking it easy. Goddard and I went to *Chang Tang yang* Air strip. Got 3 letters fr Evelyn 17-28 Feb. & March 10. Pepped me up. wonderful. Got 4 carriers, canteen for squad. 6 toothbrushes, shoes and Everything. Tired as Hell climbing hills. Saw Peters, Pongrata, Chambles, Maddry from Bat. 7-3 guard duty. Tired As Hell.

April 14. Hit the Deck very Early. Mahmood & I went Down and got water then made patrol up to Sgtt Olivers. Gorretti came by and tried to Fix Radio. Mahmood went Down to Air Strip To get supplies for us. Rumored we are to get reld Today. This outpost is Hell. Didn't get Reld. got ground coffee, tea, sugar etc. Recd 4 Letters fr Evelyn. 1 from Kat &1 fr Howers. was grand. Couldn't Be Better. God is Kind. Good Rat cheese, eggs, coffee. Went to B BN Hill. Will be reld tomorrow. Swell. Tired and Sore & Dirty. Bed 8PM. Guard 10-12 PM. God please guide & protect us thru this for Evelyn's sake in Jesus' Name Amen.

April 15. Hit the Deck 5 AM. Had trouble w/ McBride. He turned chicken & wouldn't Fight. Ate Same old stuff. Cheese, Bouillon, Crackers. Guns blasted Loose. Expect to be relieved Today. God, I hope so. This is getting on my Nerves. Got Reld by Red. C. T. at 1050. Moved up to maggot Hill was put in perimeter by Sgt Mulligan. No good. Made good bunk w/ Goddard. Got water. Met good friends from A BN C. Co. Ate chow. Saw Doeling. Bed Early. Art opened up. Lots of Hand Grenades.

April 16. Up 830. Nice sleep. Quite a bit of excitement Last Nite. Fired 4 rounds. Lots of Rumors. Saw Hickman, Peters and the rest of C Co. men. They certainly like me. Cut the squad's hair and washed up. Got a Little Rest. Think we Have the Nips on the Run. Thank you, God for guiding and protecting us. Keep Them safe at Home. Dug comm Trenches. Bed 7 pm on guard 2nd. Still Lots of firing.

April 17. Up 7 Still Nice sleep. Went on patrol out in front & pick up Jap helmet, cartridge belt, Bayonet and Scabbard, canteen and mess kit. Cleaned them up & disarmed Hand grenade and Knee mortar. Washed up.

Got 10-1 Rat. Saw Mason caught Hell from That 8 Ball Mulligan. Wrote v mail to Evelyn. Swell to hear from Her. Bed early. Have Fever Tired.

April 18. Hit the Deck 8AM. good Breakfast. Went to 3rd plt Am going to see Capt. Haol. Watch fixed by Blue. Were issued mountain Rations. Swell. Washed & shaved. Wrote letter to Evelyn. Am full. Hope to go to Co HQ. Bed Early. Saw Captain.

April 19. Hit the Deck. Ate Dried Beans & coffee. Not Bad cereal Smuck's squad went on patrol with red to find Hick. Orange column is going to air strip is swell. Left 11AM. ARRived 2 PM. It is wonderful to get the rest. Pongrata outfitted me in clothes. Swell. Saw Maddry. Sleep early After Bath. Mahmood transferred to 3rd plt.

April 20. Up Early. All of squad got new clothes. Was issued 10-1 Rats. Would love to see my Honey. God thank you for carrying us safely thru. please God continue. Rhumors Everywhere. Don't know what we are going to do. Nice chow. Was also issued J. Rats. Nice took bath. Made Lt. Smaules some candy. Bed Early.

April 21. Friday. Up 5AM. Reveille 6AM. 2 miles fr Japs and Behind their lines. Phooey. 1 hour of close order drill & then BAR then full field inspection, But it Rained us out. Got wet as Hell. Saw Pongrata. Got Rations C. Chinks come in tomorrow. We are moving out tomorrow. 1 ___ Bed Early.

April 22, Sat. Hit the Deck. Have very bad chest cold. Lots of chow. Got some Sulpha diazine Tablets. Will move at 1PM 800 yards. Chinks to take over. Am sick &

tired. God please Help us, guide us safely & protect us thru the coming storm. Money moved to new ARea. No perimeter. Nice area. Mac & I got water. Saw old pals from other plt. Bed At 830.

April 23. Sunday. Up 6AM. 1 man water guard. good chow. Bacon, roast beef. Chinks moved in area. Goddard was sick as Hell Last Nite. Got cigts traded for chink, Jap and India & Ceylon, Money. Clark and I got __ for plt. Rested good. Moving out Tomorrow.

April 24 Mond. Up early. Moved out at 8AM All up Hill 15 miles in 8 hours. Rough. Into Wellington at 5PM. Wet w/ sweat & tired. Ate chow. Had water guard. Made S/Sgt 16th of April. 300 Japs 1 days march S.E. Tied in w/ 88th Chinese Regt. Good Looking boys. Bed Early. Tired and sore.

April 25. Tuesday. Up 530. Chowed up. Slept well. Rolled pack. Are moving 7 miles. up Hill is rough. Leave out at 9AM. Rain &sweat. 4 miles up, 3 Down.

Chinese in front A BN. Taylor is a damn sponger. Wet & tired. Bivouaced. Ate good. Bed Early. Slept well.

April 26 Wed. Up Early Ate K Rtns & ready to move At 8. Going from 1 to 4 miles. Very Hilly to Nabum. Divisional HQS. The 3 miles and ARRived at 10AM. 1st Sqd was point. Bivouaced. Took Bath. Went on patrol. Found Nip Trail. Was Rough Mac got Fish. We cooked in bamboo. Bed 8. Tired & sweaty.

April 27. Thurs Was Trfd to 1st plt. Whole SQd Swell to leave Mulligan & Smith. Moved out at 830. Up a

Rough Hill. Hot as Hell. 4 miles in to bivouac. Took bath, washed clothes & read Omnibook. Smoked pipe. Drew 2 days Rat. Mac & Snyder made candy. Bed Early. Built Banana Shelter.

April 28 Friday. Up 5 AM. Rolled packs. Moved out to Nabum then E. 9 miles in All. Saw all old Boys from A BN. Hot as Hell. Went up Hill. 7 miles. Getting Rougher. Wore out. Moved into bivouac area at 530. Had 1 drink believe it or not. Made candy. Rained. Bug Bite all Night.

April 29 Sat Up cold as Hell. Clark and I got water out of mud Hole. Ate breakfast. Moved out at 8AM. Marched up Hill 3 ½ miles. Boy its rough. Rained During Dinner break. Miserable. Lost 12 Animals from Landslides. Bivouaced At Edge of Trail. Chinks Behind. Khaki in front. Pack is Lighter. Slept w/ Goddard.

April 30 Sun. Up and at em. It rained 7-830. up Hill mud & slippery. Had to Dig out Trail. Moved 4 miles. 1st plt is in the Lead. Just one more Big Hill After This one. Made dextro candy good. 14 Horses went over side of Hill. Bed 730. Rained All Night.

May 1st Mond. Up 530. At Breakfast. Rolled pack. Moving out 730 Reached Top of Hill 10AM. Boy it was wonderful to walk Down Again. Muddy as Hell. Rained all Day. Dug New Trails All Day. Rough Reached Bottom 4PM Bivouaced. Nice Supper. Bed Early Muddy & Tired.

May 2 Teus up 5AM Ate Frugal meal. Mac went to sleep on guard. Its bad. Have 6 miles of Bad Trail today.

Muddy & All Marched 10 miles up and Down. We were rear guard. Arrived 5pm. Slept on side of mt. Kachins contacted nips.

May 3 Wed. Up. Sick call at 7. May stay Here Today. Have Air Drop. Slept well. Leaches are bad. Went to within 300 yds of Nips. Kachin scouts came Back and put up outpost 1500 yds Down River. No food for 2 days. Ate fish & rice. 2 man guard tonight. Bed Early.

May 4 Thurs up Early moved to Top of Hill. Drew 4 Days Rations. Moved out at 9 AM Followed Chinese Marched 10 miles. passed Khaki. This is a Big Drive Chinks, British ghurkas, Indians Kachins to Mishinau Rough. Made Candy Bed 8AM (PM)

May 5 Fri Up at 5:30 Raining. Left out at 8AM 4th squad is point. Expect Japs. Can't tell Marched 6 miles. Muddy tired & sore. Ate some rice. Not bad. Ate rice for Dinner waited for I & R report on Nips. Bivouaced At Hill. Bed 8:30. After talking to Kasuly.

May 6 Sat. Up early Strout went to sleep on guard. Bad. Men are wore out. Leaving out Early. Moved out at 7AM. Break & march contacted Nips. Flanking them. Moved 6 miles. Bed After Dark. Rained.

May 7 Sund Up 5AM Ate & Taylor moved w/Radio. Rain and wet. Marched 5 miles. Rained. Squad stole extra rat & tea Swell. Nips are firing to Beat Hell & we Bivouaced. Bed Early guns All Nite.

May 8 Mond Up 5AM cleaned guns. Firing All ARound us. 1 ½ hour guard. Ate good meals. Air Drop

Today Layed around. 1st Bn is pulling ahead. Good.

Another Chinese Regt pulling in. God please guide us safely thru this Jesus

Name Amen.

May 9 Teus Up Early. Miss my honey very much. May move out soon. c rats out. Goddard went on Detail. Chinks attacked Japs. No outcome yet. Japs lost 17. Chinks 26. We move on Down main Trail. Hot as Hell. Bivouaced 430 Good sleep. Still fighting Behind us.

May 10 Wed Up 5 a.m. Rolled packs, Ate, got water. Still firing. Made candy. Have listening post. A Bn passed through us. Its About Time. Chinese swiped Rations. Expect to move out tomorrow.

May 11 Thurs Up early will move out this morn A & Chinese are in Front on Another Trail Left out 9:15 marched 9 miles. Bivouaced at 4PM. Took bath. feel Better. God Be with us. Made candy. Monsoon coming. Tired and sore.

May 12 Fri Up 5 a.m. Ate 2 rats Move out at 8AM getting closer to Nips. Contacted Them close to Noon. 2 dead. Noe in 1st plt & 3 wounded. Bad Business. Dug in. I got 1 or 2 of them. No sleep. God please be with us. God in Heaven guide us safely thru The Bad Day Tomorrow & please protect us for Evelyn's sake and in Jesus name. Amen.

May 13 Sat Up 5 AM Very little sleep last night. Got Another 1 of our men bad. Bullet hit my shoulder. God please guide us safely thru this and protect us for

Evelyn's sake and in Jesus name. Amen. Hit nips Again. Was trapped for about 10 min Rough. God be with us in our Hour of need. Nips have cut our back trail off. Bad spot.

May 14, Sund God please guide & protect us. Carry us thru this safely. Are withdrawing. 1st plt is Rear guard. Rough. God got us thru for Evelyn's sake & IN Jesus name Amen. Marched 8 miles into area at 6 pm. Chow and Bed. Bread & jam.

May 15, Mond Up 5am Ate cooked candy. Took Bath. Have Bad cold. Chest & Head. Monsoon has started. Left out at 9 AM marched 12 slippery & muddy miles. In at Dark. Chest cold is Bad. Sleep 10 pm.

May 16, Teus Up early 5 AM Left out 6:30. Raining. Up muddy hill. Pvt. Goddard. Covered 11 miles. Rough & slippery. Made candy At Noon. Chest is worse. Am The point into bivouac Area. Cross 4 rivers. Bed early. T & S my chest.

May 17, Wed. Up early. Moved out 7 am. Khaki Leads. Marched All Day. Rained. Good Level Trails. Marched 12 miles. Swiped 10-1 Rations. Swell. Moved out at 7 pm. Marched Another 9 miles. Tired & sore. Close to Mishina our objection. A Bn is there. Sleep 10 pm.

May 18 Thurs Up 4 am. Fixed packs moved out at 5:30. Held up till 7:30 pm Coffee while you wait, Dive Bombers Raising Hell. Troop transports Landing. Marched 10 miles into Mishkina. Few Japs Held us up. 8 were killed. Got SunTan. Dried & cleaned up. Nips are just a short Distance away. Bed Early. God guide & protect us. Amen.

May 19 Fri up 5:30 Rained us out about 5 AM Got codine fr Doc. Chest is in a bad way. Coughing up blood. 500 Nips Headed this way from North. No Rats for 3 days. God in Heaven guide us safely thru this & protect us for Evelyn's sake in Jesus name Amen. Got Rats late. 5:30 put outpost 800 yds out. Cont patrols. Bed early. Expect nips.

May 20 Sat Rained us out Again. Am wet & cold. Chinese troops came in on planes. Big battle ahead of us. British crossed Irrawaddy. Nips near outpost. Rained til 1220 Was ___ Moved up road to Another position At 1pm. Sun is hot. Dug in BAR 100 yd Apart. Ate Bed early. Lots of mosquitos.

May 21 Sund Myitkina up Early. Packed up. Rained like hell. Attacked Down Road to Myitkina Got hit by nips at road Junction. Bullets Everwhere. God guide us safely thru this & protect us in Jesus name Amen & for Evelyn's sake. Amen. Fired all day. Rained wet & sore. Pulled back & built peremiter 75's Hit Nips. Lost m.g. was bayoneted to death. Took m.g.

May 22 Mond Up 4AM. Rough Last Night. Mosquitoes, bullets, art. mortar. Dug in Deep. No food Again. 3 days Before. Rolled packs. Raining. Got food AT 9:30 Moved out at 10:00 AM. Set up old positions in village. Have a hell of a HeadAche.

May 23, Teus Rained All Night. Men's nerves are Breaking. Trigger Happy. 14 men being evacuated.

Typhus. Cold & wet. No news. Sicker than hell. 102.8 Fever Typhus. Slept all eve. Saw A Bn Bed Early. Same place.

May 24 Wed. Up all night. Nips Hit us Last Night & Raised Hell 920. Inside of perimeter. Killed Lt. Smith & Hogan, Sgt. Mattlock, Shafer & Burton. Bayoneted & shot 7 more men. No sleep. Fired All Night. Nips All Around. Rain continuous. Moved out at 11AM Hit Mogong Rail Head. Set up perimeter. I was given mt slip and went to Airstrip. Ate Then. Was picked up by troop carrier. 1 hour 15 min. to Ledo. Slept in ward. Is wonderful. No nips close. Have blue spots over me. Spitting up Blood & Shrapnel wound.

May 25 Thurs up early. Swell breakfast. Maner & I caught ambulance to Marguerite ward. B.17. Took swell shower. Went to the Red Cross toilet. Out & then back to bed. Doc Looked me over. Looks bad. No mail nothing. Still spitting up blood. Black spots. Bed early.

May 26 Fri. Up at 7:30. Raining. Still in Margerite General Hosp. 4 docs came around. 4 looked me over. Bad. Feel worse. Good chow. 3 more docs came in to look me over. Bad case. Wrote Evelyn. Good chow. Bed 9:30.

May 27 Sat Up 7 am. Good breakfast. 2 docs this morn. Read books. Not much else to do. Tired of Laying around. Feel bad. Bed early. Won 25 Ruppees.

May 28 Sund Up early. Made bed. Another blood Test. 3 times they have stuck me. Rough. Good meals. Sun is shining swell. 10 new cgts Socks & drawers. No shoes. Won 31 Rupees. Bed late. Saw lot of old friends. God guide us safely & protect us on our way for Evelyn's sake in Jesus name. Amen.

May 29 Mond (Monday. Up early. Still nothing to Do. Laying ARound. Its rough. Wrote to Evelyn, Atchinson, Ida. Tired as Hell. Read 3 books. Good chow. Won 23 rupees. Poker.

May 30- Teus Up Early. Had to go before over 100 docs for Strange New case. Don't know what it is. Tried to get Barracks Bag. Lost 3 R poker. God carry us thru safely & get us back to the States safely for Evelyn's sake & in Jesus name. Amen.

May 31 Wed. Up Early feel like Hell. Docs looked me over. Young brought in 30 R worth of candy and other things. Won 8 R in card game. Hot as Hell. Bed Early. Letters to Mama.

June 1 Thurs Up early. Got haircut. Had watch fixed. Good meals. Pulled Hot Hell. Wrote Evelyn. Sent Jap & Chinese money. Bed late. Took shower. Saw Westy & Doeling.

June 2. Fri Up early good breakfast. Went to get xray. Saw Joe Diskin Good boy. Played poker. Lost 22 Rupees. Bed early. Read books.

June 3, 44 Sat Up & at em. Feel Better this morn a11 of 2 Bn is now out of Myitkina good. Saw Westinghouse. Hot. I took xray. Saw Joe Diskin is going into Myitkyina.

June 4 Sund Up early not much doing. Good meals got all my clothes. Won 45 Rupees. Saw same old friends. Bed early. No letters. Wrote Evelyn & others.

June 5 Mond Hit the deck early. Good breakfast. Low on Cgts. More patients come in more go out. Layed in bed all day. feel bad. Played cards. Lost 40 Rupees. Bed early.

June 6 Teus Same as yesterday all but I won in poker. 40 some odd Rupees. Went over to guard house. Cleaned 25 caliber rifle & oiled it. Not much doing went to staging area. Drank beer. Bought____. Rode in on time Saw Maddry & old friends. Bed early.

June7 Wed. Not nothing today. Doc came in. new one.. 97 new nurses came in. Feel good. Raining. Went to 14th evac & drank beer. Saw old friends. Heard Bevon and Granfield was in Myitkyina. Back & in bed early.

Thurs June 8. Up and went to mess Hall for chow. Drew picture. Doc will fix me up. Feel pretty rough. Not much doing. Saw Doeling. Drank beer. Feel pretty bad. Lost 10 Rupees.

Fri June 9 Up early feel fair. Not much doing. Is Hot as Hell. Got 24 letters. from Evelyn, Ida. Mama, Papa, Buddy, Helen, Inez. All of them. J Redly ans his letter and Mama. Bed late.

Sat June 10 Still nothing to do. Hot. Raining. Good. Rumor am going to States. I hope & pray to God that it does Happen.

Sunday June 11. All the boys got paid today. Big poker game so it seems. Saw Doeling. He is going to14th

Evac. New nurse. Its nice there.

Monday June 12. Still in 20th Gen. Not much doing. Read. Layed around. Feel Bad. Went to Lab & had blood count & clotting. Red Cross for cgts.

Teus June 13-44. Still nothing. Letters from Evelyn & Mama. Swell. Ans them. Slept. Pvt _

Wed. June 14. Up & at em. Feel better. Went to show _____ was in Burlesque. Swell Band. Layed around. Nurse nice.

Thurs. June 15. Hit the Deck early on ward. __went with Sharkey & Paddock.

Drank beer. Swell in Late. Saw good show. Miracle of Morgan Creek.

Fri. June 16 Up early. Went to chow. Cleaned up. Moved us to ward B9. Saw show. Very good.

Sat. June 17 Not much doing Up 7 at em. Layed around. No man gone. Some for my honey. Bed Early.

Sun. June 18. Same as yesterday. Not nothing doing.

Mond June 19 Feel pretty bad today. Fever 101 Blue spots. It's a relapse. Lots of docs.

Tues June 20 Worse than ever. F.102.6. Spitting up blood. Gosh feel rougher than Hell. 9 letters from Evelyn. Very sweet.

Wed. June 21. Worse. Still more blood & f 103.6 Getting worse. God please Help me. In Jesus name

Amen.

Thurs. Jun 22. God I'm getting worse. More blood spitting up f 104. Will get blood trans tomorrow. Weak as hell.

Fri. June23 Had blood trans today 1150 cc of it. Fever 105.6. Ice packs all day. Feel rough. No more spots. No more blood.

Sat June 24 Feel better today. Fever lower. No blood come up. Ice packs okay I guess.

Sun June 25. Feel better today. No fever. No blood up. Can't eat & very weak.

Mon June 26. Am snapping out of it. Friends came in. Not much to eat. Still weak as the Devil. Letters from Evelyn. Swell to hear from her. God is kind.

Teus June 27. Wrote Evelyn & Mama today. Feel a lot better. God please carry me back to States soon & safely for Evelyn's sake in Jesus name. Amen.

Wed. June 28. Up & at em. Shower. 1st in two weeks. Feel lot better. Wrote Inez & Ed. God is Kind.

Thurs. June 29. Up early feel quite a bit better. Read good books. Swell letters from Evelyn. 3 of them.

Fri. June 30. Again up at 6 am Not much doing wrote Evelyn and ____ others. Good meals. Talked to ____ from La. and C BN Boys.

Sat. July 1 No mail today. Gosh I miss the letters. But God is in charge. Bless His Name Feel quite a bit better today Bed very early couldn't sleep. Stanley brought fruit & Juices.

Sund July. 2 Still nothing doing today Can't go to church no mail slept & Read

Mon July 3 woke up at 6AM Read played chess Saw Doeling some news letters from Frank, Ida, & Evelyn. Swell Ans. them.

Teus July 4th Darn ward Boy Woke me 545 Too Early its getting Hotter than Hell All 2 yr. men are going Home. Saw Baker & Kramer good meals. Bed Early

Wed. July 5. still nothing Doing No letters spent last 2 R.s on watch Its fixed now. Went to staging to find Bag Saw Graham & lots of men my friends No Go. Bag Lost.

Thurs July 6 Oh this darn monsoon rain All 24 hours guess I'm in Det. of Pats. God please guide me safely back to Evelyn for her sake & in Jesus name Amen.

Fri July 7 Chess, cards, & reading That's all I do. Good chow. I signed the pay roll Now I am the Det of Patients

Sat July 8 Up & about still very weak Saw show Mark Twain Schoemer and I was good. Feel bad.

Sund July 9 Not much to do today. Read Book. Saw Goddard. Good Boy. Played cards & went to show

Monday July 10 Up no letters. Just laid around Saw good show. Read quite a few books. Good card games.

Tues July 11 Schanheur went to Ledo brought back

pie. Oh Boy. Some more fruit juices from Stanley. Letter from Frank.

Wed. July 12 Everything will turn out for the best. God help us. Good news 2-year men going home maybe me too.

Thurs July 13 Saw show Around the World ___ Darn good. Read books.

Fri July 14 No news today. I love my Darling. God listened to my prayers Amen

Sat July 15, Not much to do today Might go to States. Joe Disk came in Looks good— is in B7 Nut ward. Bed early.

Sunday July 16 Up & at Em. I go to regular mess Feel good. Went to staging area. Saw girl & Freeman 4 letters fr Evelyn 1 fr Mama & Ida. Swell.

Mond July 17 Wrote 6 letters to friends then to staging area. Saw pretty red cross gal. Feel better. No letter.

Teus July 18 Not much doing today Hot as Hell no letters Bed Early

Wed July 19 Same as yesterday but I wrote Home & to Evelyn. Love her very much.

Thurs July 20 Miss Evelyn much

Fri July 21 Went to 14th evac found Bag. Swell Letters from home.

Sat. July 22 Still not much doing Hot as the devil.

Sweat much.

Sund July 23. Still hot got new clothes no news Wrote ___. God Amen.

Mond July 24 Good news Going Home Fri. God ans our prayers. Thank God. Amen Letters from Evelyn & Buddy, Lloyd & Maime ans them

Tues July 25 Up and at em. Saw good show. Am going Friday. I Hope back. God be with us. Amen

Wed. July 26 Same today. Playing cards & reading books ___ had shirts cut down.

Thurs July 27 restless today. Saw Schanheur & ___ Saw Stanley Good bye & Thank God for all he has done for me.

Fri July 28. On Plane at 8 AM At Ledo Flew to Goya C47 changed to C45 21 of us and on to Agra. Miss Jolend & rest. Nice nurse. Tired. Bed at 2PM

Sat. July 29 on plane at 630 off 7AM Arrived in Karachi at 12 Noon Bed 1030 to 81 Gen Hosp swell ward took 2 showers & drank cold cokes 1st in 2 months & ice cream. God I pray be with us & carry us safely to States for Evelyn & mama's sake in Jesus Name amen.

Sund July 30 Up at 730 good mess. Nice nurse won 2 R.s in poker went to Karachi at 1PM no good. Ate ice cream Got bag for Evelyn & moonstone & had picture taken. Back to camp by taxi Bed & Tired.

Mond July 31 Up & at em Expect to move out today or tonight at 4AM. Played poker lost, Packed 25 Cal.

Rifle in Bag.

Tues Aug 1st. Up early read magazines, 8 letters from Evelyn & wrote 1. Played poker won 6 Rupees. Spent $12 in Karachi. Sun & saw Ann Sherriden

& Bon Blue yesterday 5pm. Went to show was called at 9P.M. Left Karachi t 12pm Midnight by C54

Wed Aug 2 Left Karachi 12 midnight Arrived in Aradia. Had breakfast. Ate dinner on plane. Flew on to Egypt near Suez canal got 7 hours of rest Went to PX good meals. Left at 7PM Flew all night to Gold Coast Africa.

Thurs Aug 3 Got in at 9AM for 6 hours rest. Nice place Shave & cleaned up. Nice meals. Slept & had cokes. Left at 11AM Arrived at Assention Island at 6 PM. Good meal Up & flew to Natal Arrived 12:30AM to bed Natal

Fri Aug 4 Nice sleep Up 6AM 50 cent Breakfast on plane at 7AM Few to British Guinia 2:30 PM I was too sick to eat on plane at 5PM Left Guinia for Porte Rico God is very Kind. Sick as Hell Arrived Puerta Rico at 9PM 1 hour rest. Ate. Left at 10PM for Miami Beach.

Sat Aug 5 Slept Arrived at Miami Beach 3:24 AM Temperture then customs checked the baggage on bus to Biltmore Hotel Hospital Swell Thank you God. Rested ate ice cream & coca colas swell Called Evelyn at 730 got call thru at 10:30 Oh Boy it was swell to hear her voice. Asked for $100.00

Sun Aug 6. Fooled around all day. Rested read

would like to see my Darling Oh Boy to bed 10pm walked around with Landrath. I watched bathing girls.

Sun Aug 7 Up Early took temp then got money order from Evelyn God socks, paid Landreth $10.00 Love to be with my darling.

Mond Aug 8 Up & at em God is kind I go to mess hall today This Biltmore is a grand old place pools, golf, tennis, ___ and everything. Sent Darling money purse & pictures. Grand Love her.

Teus Aug 9 Docs came in today Checked me over Eyes Balance ears so far so good Saw all my friends not much doing except to town with Landrath Drunk as hell. Gosh feel bad.

Wed. Aug 10 Called Evelyn tonight Wonderful woman God is kind Good meals Wrote to Jean & Ed. 7 AM rest. Like to hear from Frank & Oscar

Thurs Aug 11 feel bad this morning Bottled ___ not bad Wandered around. Went to town with pass Oh Boy good time. God I love you. Take care of Evelyn Too Drunk Broke feel bad.

Fri Aug 12 Still nothing to do fooled around good meals, Had clothes Dry Cleaned. FDR will speak tomorrow night 6pm. To bed early tired.

Sat Aug 13 Up early good breakfast cool today people like my beard. HoHo. God please carry me to Evelyn & Mama safely for their sakes and in Jesus name Amen. Will listen to Roosevelt 8 pm Not much to do bed

8:30.

Sund Aug 14 13 Up bright & early. Fooled around Would love to see my darling Went to town got about ½ drunk. Came in early about 10pm.

Mond Aug 14 Not much doing. Drank lots of malts & cokes. Went to town with Landrath met girls but didn't fool with them. Evelyn called. Beautiful devil Gosh I love her Got return on letter from J.D.

Tues Aug 15 Up & at em feel a lot better had malts & ___ Wrote 5 letters Mama, Frank, Sudie, Evelyn & others. Phoned to my darling Evelyn.

Wed. Aug 16 Up still nothing to do read western book Slept awhile Took bath Saw Phil Murray pianist play at patio ___ then to bed.

Thurs Aug 17 Up read some more books then fooled around not very much to do. Slept & then went to town.

Fri Aug 18 Still nothing to do this is driving me crazy. Around & then to show. Not bad. Then to bed. No letters Wrote Evelyn

Sat. Aug. 19 Just got word I was leaving 630 the 20th Swell helps out Got pass and went to town at 1PM Had shirts sewed & then home. 2 letters from my darling a $10.00 bill in one. Swell Phoned at 1030 swell to bed.

Sund Aug 20 Up 5AM had early breakfast then left out on flight. Left out at 730AM to Tallahase Fla. To Jackson Miss C47 is rough. They cross the Miss. At 3pm over Lake Charles at 3:15 landed in Tyler 430 then to

Dallas Overnight Had pass with swell to town & in 12 PM

Mond Aug 21 Up feel bad. On to plane at 830 left out. Long trip 2 hours to San Antonio then off again & landed at Santa Fe at 530 Swell place In Q21. Phoned Evelyn & she will be here on Sat. To bed rested well. Swell place went to chow.

Teus Aug 22 Up 7 AM go to chow 7AM Had blood taken. Shots and tests rough. All morning. Then took sun bath Everything Oh boy wonderful. God is kind.

Wed. Aug 23 Not much to do today, read books & magazines & then into town 3pm. Couldn't find any apartment but will try fri then I went to orchard ____ just had 1 dollar. Into Hosp at 10pm to bed.

Thurs Aug 24 Up at usual time. Read (I was a ___ flier. Very good. Slept in afternoon Swell place went to chow. Good. Called Mrs. Hanchey at 10PM Evelyn is coming out at 12PM finding no apartment.

Fri Aug 25 Up 6AM went to chow. Swell. Expect call from Evelyn 11:30 she's here. Hurray God is wonderful to answer our prayers like this. Had pass Saw my darling at 108 San Franisco St. She is wonderful. Ethel was with us.

Sat Aug 26 Up early & grand Swell to have my honey with me. Went on pass at 2PM to hotel and had dinner Tried to find apartment. It was hard. Had to buy blankets and sheets Will move tomorrow. Only God can understand how I feel with Evelyn.

Sund Aug 27 Up with my Darling At 7:30AM. Our love is larger and broader than ever. To breakfast 10:30 We went to look over lights lots of old places Moved into apartment at 5PM. Needs cleaning up. Back to Bruns General Hosp.

Monday Aug 28 Hit the floor at 7AM and then to breakfast. Tired and Sore. My honey came out from 2-4. God please give us a sick leave to visit Phoenix for Evelyn's and Mama's sake in Jesus' name Amen.

Tuesday Aug 29 I am very much enlightened. I am getting a discharge from the Army and God will only answer our prayers a good compensation please give me what you can for mine and Evelyn's sake. Amen Took Sgt H___ to see Evelyn. Grand.

Wed. Aug 30 It won't be long now. The fiesta was Sat & Sun. Had a big watermelon. Oh Boy ____ Wrote some letters home. Wonderful stuff. God I love her.

Thurs. Aug 31 Not much to do today. Played cards read. Will have operation tomorrow. Went home had couple drinks & hot tamales.

Fri. Sept 1 Four years ago & ____ going into ___ 16th. Had operation. God it hurts. Went home played poker. Lost 15 cents. Had melons not bad.

Sat Sept 2 Went to bungalow. Went to Fiesta Got drunk as devil. Danced and hurt. Had good time. In at

12 pm.

Sun Sept 3 Up 9AM. Evelyn had breakfast fixed swell. Up and about. Good. Washing rough. Slept and went to town. Ate in 8:30. Ate Watermelons, hot tamales. Played poker.

Mond Aug 4. I go before the board. Good meals. Not much doing. Passed C.D.D. Feel bad about it. Will go out Thurs. Today went home up until 7:30.

Tues Sept 5 Not much to do. Read and straightened out Service records oh boy it was rough. Evelyn came out. Love her.

Wed. Sept 6 Time is drawing closer. Gosh one more day and I will be Mr (Mister). Evelyn came out. Love her. Maybe Junior got clearance. Wrote Frankie and Landrath.

Thurs. Sept 7 Up early. Feel funny.

Works Cited

1. Cartwright, Mark. "Black Hole of Calcutta." World History Encyclopedia., October 11, 2022. https://www.worldhistory.org/Black_Hole_of_Calcutta/.

2. Michael Gabbett, *The Bastards of Burma: Merrill's Marauders and the Mars Task Force Revisited*. (Albuquerque, New Mexico: Desert Dreams Publisher. 1989), xiv.

3. James E.T Hopkins and John M. Jones, *Spearhead: A Complete History of Merrill's Marauder Rangers*. 2nd Edition. (New York: Merrill's Marauders Association, Inc. 2013, xix.

4. Charlton Ogburn, *The Marauders*. (New York: Quill.1982), 11,16.

5. Ogburn et al, *The Marauders*. 16.

6. Military Intelligence Division U. S. War Department, Merrill's Marauders. (February-May 1944. American Forces in Action Series. (Washington D.C.: 4 June 1945), 114.

7. The birth of the Fighter Plane, 1915. Accessed 2022. http://www.eyewitnesstohistory.com/fokker.htm.

8. Lurline Brochure, Author's personal collection

9. Ogburn et.al., *The Marauders*, 29.

10. Ogburn et.al., *The Marauders*, 30.

11. University, Carnegie Mellon. "Historian Examines Japan's Unexpected Alliance with Nazi Germany" Carnegie Mellon University,n.d. https://www.cmu.edu/dietrich/history/news/2019/law-book.html.

12. Col. Logan Weston, *The Fighting Preacher*. (Alexander, North Carolina: Mountain Church, 2001). 103.

13. Gabbett et.al., *The Bastards of Burma*. 29.

14. Ogburn et al, *The Marauders*, 69-70.

15. Hopkins and Jones et.al., *Spearhead: A Complete History of Merrill's Marauder Rangers*.111.

16. Pletcher, Kenneth, *Bushido Japanese: "Way of the Warrior"* https://britannica.com/topic/Bushido.

17. Ricks, Gregory. "Army Mules: The Beast of Burden in War." Warfare History Network. Sovereign Media, July 28, 2016. https://warfarehistorynetwork.com/ army-mules-the-beast-of-burden-in-war/.

18. Gurung, Tim I. "A Brief History of the Gurkha's Knife – the Kukri." Asia Times. Asia Times, February 18, 2020. https://asiatimes.com/2018/04/brief-history-gurkhas-knife-kukri/.

19. Hopkins and Jones et. al., *Spearhead: A Complete History of Merrill's Marauder Rangers*, 145.

20. Hopkins and Jones et. al., *Spearhead: A Complete History of Merrill's Marauder Rangers*, 144

21. Hopkins and Jones et. al., *Spearhead: A Complete History of Merrill's Marauder Rangers*, 174-175.

22. Lt. Col. John George, *Shots Fired in Anger*. (Washington D.C.: National Rifle Association of America) 485-486.

23. Hopkins and Jones et. al., *Spearhead: A Complete History of Merrill's Marauder Rangers*, 191.

24. Hopkins and Jones et. al., *Spearhead: A Complete History of Merrill's Marauder Rangers*, 188.

25. Lt. Col. John George, *Shots Fired in Anger*. (Washington D.C.: National Rifle Association of America) 483.

26. Hopkins and Jones et.al., *Spearhead: A Complete History of Merrill's Marauder Rangers*, 138.

27. Colonel Charles N. Hunter, *Galahad*. (San Antonio, Texas: The Naylor Company. 1963) 45.

28. National Toxicology Program Executive Summary of Safety and Toxicity information Halazone Oct. 4, 1991. https://ntp.niehs.nih.gov/ntp/htdocs/chem_background/exsumpdf/halazone_508.pdf

29. Col. Logan Weston, *The Fighting Preacher*. (Cheyenne, Wyoming: The Vision Press.1992), 142-143.

30. Hopkins and Jones et. al., *Spearhead: A Complete History of Merrill's Marauder Rangers*, 316.

31. Hopkins and Jones et. al., *Spearhead: A Complete History of Merrill's Marauder Rangers*, 321.

32. "Military Intelligence Service Language School." Military Intelligence Service Language School | Densho Encyclopedia. Accessed 2020. https://encyclopedia.densho.org/Military_Intelligence_Service_Language_School/.

33. Hopkins and Jones et.al., *Spearhead: A Complete History of Merrill's Marauder Rangers*, 330.

34. Military Intelligence Division U. S. War Department, Merrill's Marauders (February-May 1944. *American Forces in Action Series*. (Washington D.C.: 4 June 1945), 91.

35. Weston et.al. *The Fighting Preacher*. 273.

36. Viktor E. Frankl, *Man's Search for Meaning*. (Boston: Beacon Press) 177-178.

37. Hopkins and Jones et.al., *Spearhead: A Complete History of Merrill's Marauder Rangers*. 478.

38. United States of America Office of Price Administration, *War Rations Book No 3* (U.S. Government Printing Office:1943).

39. World Health Organization. https://www.who.int/csr/resources/publications/surveillance/plague.pd (2017 October 31).

40. "Medicine." Medical Developments during WWII, https://medicalwwii.weebly.com/medicine.html

41. Ibid

42. "D-Day Invasion Was Bolstered by UW–Madison Penicillin Project." News, 2 June 2017, https://news.wisc.edu/d-day-invasion-wasbolstered-by-uw-madison-penicillin-project/.

43. Hopkins and Jones et.al., *Spearhead: A Complete History of Merrill's Marauder Rangers*, 306.

44. Edward A. Rock St. *The Burman News*, (Merrill's Marauders Association, Inc.) June-August 2008. 2-3.

45. Military Intelligence Division U. S. War Department, Merrill's Marauders. (February-May 1944. *American Forces in Action Series.* (Washington D.C.: 4 June 1945), 113.

46. L'Amour, Louis. *Mojave Crossing.* New York: Bantam, 1989.

47. Hopkins and Jones et.al., Spearhead: *A Complete History of Merrill's Marauder Rangers,* 676.

48. Einstein, Albert. *Letter to Morris Raphael Cohen.* Letter. Delivery date (March 19, 1940).

49. "SCENES FROM CHINESE-U.S. OFFENSIVE TO OPEN LEDO ROAD." CBI Roundup - June 29, 1944 - china-burma-india theater of World War II. Carl Warren Weidenburner. Accessed February 8, 2023. https://www.cbi-theater.com/roundup/roundup062944.html.

50. Merrill's Marauders Veterans to Receive Congressional Gold Medal September 23, 2020, https://apnews.com

51. 116th Congress, Public Law 116 – 170. October 17, 2020, https://www.congress.gov/116/plaws/publ170/PLAW-116publ170.pdf.

About the Author

Linda S. Cunningham was raised in Lake Charles, Louisiana, by Larry and Evelyn Stephenson. She worked for over twenty years with the Merrill's Marauders Association and the Merrill's Marauders Proud Descendants Association to petition Congressmen and Senators for the cause of the Congressional Gold Medal for the Marauders.

She previously wrote and published a historical fiction novel *Early Thursday* which chronicles a fictional family set during the infamous Hurricane Audrey that ravaged Cameron and Lake Charles, Louisiana, in 1957.

Mrs. Cunningham has a B.A. in English and Creative Writing from the University of Houston. She also has a B.S. and M.Ed. in Health and Physical Education from McNeese State University in Lake Charles, Louisiana.

She has won awards for screenwriting.

Author Website:

https://www.lindascunningham.com

Made in the USA
Columbia, SC
12 December 2023